AF352854

Computational Fluid Dynamics (CFD) of Chemical Processes

Computational Fluid Dynamics (CFD) of Chemical Processes

Editor

Young-Il Lim

MDPI • Basel • Beijing • Wuhan • Barcelona • Belgrade • Manchester • Tokyo • Cluj • Tianjin

Editor
Young-Il Lim
Hankyong National University
Korea

Editorial Office
MDPI
St. Alban-Anlage 66
4052 Basel, Switzerland

This is a reprint of articles from the Special Issue published online in the open access journal *ChemEngineering* (ISSN 2305-7084) (available at: https://www.mdpi.com/journal/ChemEngineering/special_issues/CFD_Chemical).

For citation purposes, cite each article independently as indicated on the article page online and as indicated below:

LastName, A.A.; LastName, B.B.; LastName, C.C. Article Title. *Journal Name* **Year**, *Volume Number*, Page Range.

ISBN 978-3-03943-933-1 (Hbk)
ISBN 978-3-03943-934-8 (PDF)

Contents

About the Editor

Young-Il Lim, Professor at Hankyong National University. His research field is process systems engineering (PSE), specific to computational fluid dynamics (CFD), techno-economic analysis (TEA), and machine learning (ML) for sustainable process development (http://cospe.hknu.ac.kr).

Preface to "Computational Fluid Dynamics (CFD) of Chemical Processes"

The rise in computational capacity has allowed improved modeling and simulation capabilities for chemical processes. Computational fluid dynamics (CFD) is a useful tool for studying the performance of a process following geometrical and operational modifications. CFD is suitable for identifying hydrodynamics within processes with complex geometries in which chemical reactions and heat and mass transfers occur. CFD has received much attention from researchers in recent years, with a two-fold increase in the number of publications in 2018 compared with 2011 (ScienceDirect, 2019). This Special Issue is mainly focused on multiphase, multiphysics, and multiscale CFD simulations applied to chemical and biological processes.

Young-Il Lim
Editor

 *chemengineering*

Review

Multiscale Eulerian CFD of Chemical Processes: A Review

Son Ich Ngo and **Young-Il Lim** *

Center of Sustainable Process Engineering (CoSPE), Department of Chemical Engineering, Hankyong National University, Jungang-ro 327, Anseong-si 17579, Korea; ngoichson@hknu.ac.kr
* Correspondence: limyi@hknu.ac.kr; Tel.: +82-31-670-5207; Fax: +82-31-670-5209

Received: 8 February 2020; Accepted: 30 March 2020; Published: 31 March 2020

Abstract: This review covers the scope of multiscale computational fluid dynamics (CFD), laying the framework for studying hydrodynamics with and without chemical reactions in single and multiple phases regarded as continuum fluids. The molecular, coarse-grained particle, and meso-scale dynamics at the individual scale are excluded in this review. Scoping single-scale Eulerian CFD approaches, the necessity of multiscale CFD is highlighted. First, the Eulerian CFD theory, including the governing and turbulence equations, is described for single and multiple phases. The Reynolds-averaged Navier–Stokes (RANS)-based turbulence model such as the standard k-ε equation is briefly presented, which is commonly used for industrial flow conditions. Following the general CFD theories based on the first-principle laws, a multiscale CFD strategy interacting between micro- and macroscale domains is introduced. Next, the applications of single-scale CFD are presented for chemical and biological processes such as gas distributors, combustors, gas storage tanks, bioreactors, fuel cells, random- and structured-packing columns, gas-liquid bubble columns, and gas-solid and gas-liquid-solid fluidized beds. Several multiscale simulations coupled with Eulerian CFD are reported, focusing on the coupling strategy between two scales. Finally, challenges to multiscale CFD simulations are discussed. The need for experimental validation of CFD results is also presented to lay the groundwork for digital twins supported by CFD. This review culminates in conclusions and perspectives of multiscale CFD.

Keywords: computational fluid dynamics (CFD); Eulerian continuum fluid; volume of fluid (VOF); multiscale simulation; multiphase flow; multiphysics; chemical and biological processes

---

1. Introduction

Many scientific problems have intrinsically multiscale nature [1,2]. This multiscale nature often leads to multiple spatial and temporal scales that cross the boundaries of continuum and molecular levels [3]. Multiscale simulations cover quantum mechanics, molecular dynamics (MD), coarse-grained particle dynamics, and continuum mechanics [4,5]. Multiscale modeling in science and engineering couples several methods that produce the best predictions at each scale [6], which can then be applied to materials engineering [7], computational chemistry [8], systems biology [9,10], molecular biology [11], drug delivery [12], semiconductor manufacturing [13], reaction engineering [14], fluid flow [5,15], and process engineering [2,16,17].

Multiscale modeling has been used in the form of sequential and concurrent couplings [4,15]. In a sequential coupling framework, often referred to as parameter passing between scales, the macroscale model uses a small number of parameters calculated in the microscale model. The concurrent coupling strategy computes several scale models simultaneously [1,4,16]. Concurrent coupling is preferred over sequential coupling when the missing information is a function of many variables. Concurrent coupling is a powerful approach but it is also computationally prohibitive [1].

Delgado-Buscalioni et al. (2005) presented a multiscale hybrid scheme coupling MD in a subdomain with computational fluid dynamics (CFD) in a continuum fluid domain [1]. Nanofluids typically employed as heat transfer fluids include multiphysics such as drag and lift forces between liquid and nanoparticles and Brownian, thermophoretic, van der Waals, and electrostatic double-layer forces [18]. Challenge remains in nanofluid flow simulations using a CFD model that includes multiphysics and a multiscale CFD model. Tong et al. (2019) reviewed multiscale methods divided into a domain decomposition scheme and a hierarchical scheme in fluid flows with heat transfer for coupling MD with particle-based mesoscale methods such as the lattice Boltzmann method (LBM) and CFD [15]. LBM using the particle distribution function is suitable for simulating fluid flows involving interfacial dynamics and complex boundaries [15,19]. The fundamental idea of LBM is that the macroscopic dynamics of a fluid are the result of the collective behavior of many microscopic particles in the system [19]. Thus, by integrating the particle momentum space, the particle distribution function (Boltzmann transport equation) at the microscopic scale is linked to the continuous fluid dynamic properties such as density, velocity, and energy at the macroscopic scale [3,20]. LBM is an alternative approach in CFD, which has been applied to fluid flows in porous media, microfluidics, and particulate flows [19,20]. However, because CFD based on LBM focuses on the microscopic domain with complex boundary conditions, macroscopic Eulerian CFD is preferred for investigating hydrodynamics on the process level [2,15].

Lagrangian CFD approaches such as the discrete element method (DEM) and the discrete phase model (DPM) solves the Newton's second law of motion for each particle to identify the trajectories of the particles. The Lagrangian model is normally limited to a relatively small number of particles (less than 10^5 particles) because of the computational expense [17,21]. This review focuses on the Eulerian CFD for the continuum phase in chemical processes, excluding the more sophisticated methods such as the LBM and Lagrangian approaches.

Eulerian CFD has become a powerful and important tool for simulating and predicting fluid behaviors in chemical and biological processes [22]. Eulerian CFD provides the average quantities at the equipment-scale that are of practical values to engineers [23]. CFD is used for investigating the hydrodynamics of a process following geometrical and/or operational modifications [22,24]. CFD has become crucial for understanding physical phenomena in two- (2D) and three-dimensional (3D) geometries and for scaling up, optimizing, and designing chemical and biological processes [25]. However, CFD models must be validated by comparing the simulation results with the experimental data to provide meaningful information [26]. Modeling, meshing, the physical properties, and selection of a suitable turbulence model play an important role in CFD analysis.

The Eulerian CFD model is represented by the mass, momentum, and energy conservation laws described by partial differential equations (PDEs) in 2D or 3D space. As the numerical method to convert PDEs into a set of algebraic equations (AEs), the finite element method (FEM) [27,28] has been traditionally used to investigate the mechanical and structural properties of materials, whereas the finite volume method (FVM) [29], which ensures conservative fluxes within a finite volume, has been often used for fluid dynamics such as Eulerian CFD.

Fluid dynamics are governed by the Navier–Stokes (NS) equation representing the conservation of momentum. Turbulence is involved in the NS equation to consider random fluctuations of fluid motion [3]. There are several turbulence equations, such as the direct numerical simulation (DNS), large eddy simulation (LES) [30], and Reynolds-averaged Navier–Stokes (RANS)-based k-ε models. The uncertainty of RANS-based k-ε models, which originates from information loss in the Reynolds-averaging process, was discussed by Xiao and Cinnella (2019) [31].

Multiphase physics is omnipresent in both environmental and engineering flows [32]. As the precision and accuracy of manufacturing techniques progress, micro- and nanotechnologies become crucial for providing engineering solutions to problems across different industrial sectors [3]. The multiscale CFD approach has been applied to various disciplines for more accurate solutions than can be obtained using single-scale modeling. Ngo (2018) presented multiphase, multiphysics, and multiscale CFD simulations for gas-solid fluidized beds, steam methane reforming reactors, and impregnation dies for carbon fiber production [29,33]. Da Rosa and Braatz (2018) [34] proposed a multiscale CFD model for a continuous flow tubular crystallizer. The micromixing model, energy balance, and population balance equation (PBE) were coupled with the CFD model. Haghighat et al. (2018) [35] combined a 3D CFD model for smoke flow and a 1D fire event model as a far-field boundary condition to predict hydrodynamics in a road tunnel. Hohne et al. (2019) [36] combined a generalized two-phase flow boiling model with CFD to predict the breakup, coalescence, condensation, and evaporation mechanisms in a heated pipe. Uribe et al. (2019) [37] compared three different CFD models (heterogeneous micropores model, pseudo-homogenous catalyst particle model, and single-scale reactor model) in a trickle bed reactor (TBR) to show its multiphysics and multiscale nature.

The behavior of the macroscale flow is affected by microscale physical processes, which also leads to multiscale CFD modeling [3]. Multiscale behavior with complex physical phenomena, which are highly interrelated [17], appears in process engineering. For example, Figure 1 illustrates a carbon fiber (CF) production process [29,33] from a multiscale point of view. Figure 1a shows CF tape with a width of 100 mm as the final product. Fifteen tows enter the impregnation die and are impregnated with a polymer resin (see Figure 1b). The CF tape is produced from the die after it is grooved, spread, pressed, and cooled [29]. One tow having a width of 6 mm and a thickness of 130 μm is shown in Figure 1c, magnified from Figure 1b. A microscale representative element volume (REV) with 128 randomly distributed CF filaments (7 μm in diameter) is depicted in Figure 1d. A macroscale Eulerian CFD can be used for the impregnation die (Figure 1b) to investigate the hydrodynamics of the process, whereas a microscale Eulerian CFD can be applied to the creeping flow of the resin inside the tow (Figure 1d). In the concurrent coupling framework, the macro- and microscale CFD models are solved simultaneously or the two models exchange information at each time and space. If the tow domain is assumed to be a uniform porous medium, the permeability of the resin through the tow is needed in the macroscale porous CFD model. In the sequential coupling framework, the permeability computed from the microscale porous CFD model is used in the macroscale CFD model [29].

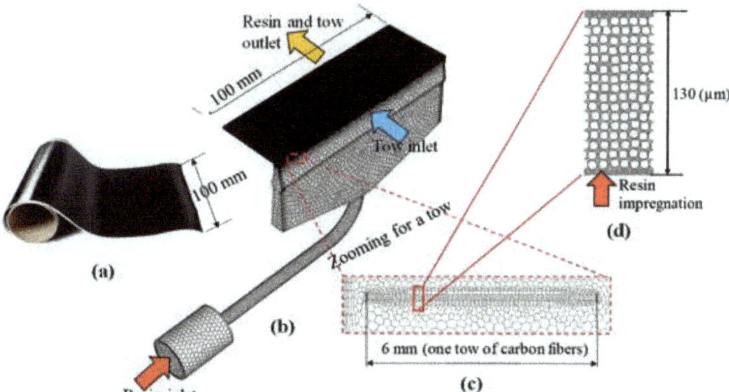

Figure 1. Impregnation die for carbon fiber (CF) tape production from a multiscale point of view (modified from Ngo et al. (2018) [33]). (**a**) CF tape; (**b**) impregnation die; (**c**) mesh structure around one tow; and (**d**) CF tow representative element volume (REV) on a microscale.

When the different models at different scales are coupled together, sequentially or simultaneously, some errors often appear at the interface where the two models meet [4]. One of the challenges is to determine how the multiscale models can be coupled smoothly. Section 3.5 shows an example of the coupling strategy.

Many reviews in the Eulerian CFD simulation have been recently reported. Ferreira et al. (2015) reviewed gas-liquid CFD simulations without electrochemical reactions in proton exchange membrane (PEM) fuel cells [38]. Karpinska and Bridgeman (2016) [39] reported CFD studies on activated sludge systems in a wastewater treatment plant (WWTP). Pan et al. (2016) presented gas-liquid-solid CFD models for fluidized-bed reactors [23]. Sharma and Kalamkar (2016) [40] used gas-phase CFD models to optimize geometrical and flow parameters that led to designing a roughened duct for the best thermohydraulic performance. Yin and Yan (2016) [41] reviewed gas-phase CFD studies on oxy-fuel combustors in a power plant. Uebel et al. (2016) reported the CFD-based multi-objective optimization of a syngas conversion reactor [42]. The interplay between electrostatics and hydrodynamics in gas-solid fluidized bed was reviewed by Fotovat et al. (2017) [43]. Pires et al. (2017) [44] addressed the recent progress in CFD modeling of photo-bioreactors for microalgae production. Bourgeois et al. (2018) reviewed CFD models in gas-filling processes, investigating the temperature evolution of the gas and the tank wall during H_2 filling [26]. Malekjani and Jafari (2018) presented recent advances in CFD simulations with heat and mass transfers of food-drying processes [45]. Pinto et al. (2018) [28] applied a CFD approach to physical vapor deposition (PVD) coating processes. Ge et al. (2019) reviewed the general features of multiscale structures in particle-fluid systems [17]. Mahian et al. (2019) [5,18] reported recent advances in modeling and simulation of nanofluids which are a mixture of a common liquid and solid particles less than 100 nm in size, focusing on 3D Eulerian CFD models for thermal systems [5]. Drikakis et al. (2019) presented the application of CFD to the energy field [3], integrated together with molecular dynamics and LBMs. Lu et al. (2019) [46] and Wang (2020) [47] reviewed Eulerian CFD approaches for dense gas-solid flows, providing a concise introduction to multiscale methods and highlighting the effects of mesoscale structures (gas bubbles and particle clusters) on the gas-solid Eulerian CFD model, focusing on the energy minimization multi-scale (EMMS) method.

Nevertheless, few researchers have reviewed Eulerian CFD studies applied to micro- and macroscale hydrodynamic problems coupled with each other. This review covers the broad scope of CFD for chemical and biological processes. In particular, it focuses on (i) multiscale CFD studies in the past decade, (ii) continuous phase rather than discrete phase, which can be formulated by Eulerian and volume of fluid (VOF) models, (iii) incompressible fluids rather than compressible fluids, and (iv) applications to chemical and biological processes. In this review, multiscale CFD encompasses a physical domain described by a continuum, excluding molecular, coarse-grained particles, and mesoscale particle dynamics at the individual scale. Eulerian CFD models in single and multi-phases are first presented, followed by applications of the Eulerian CFD to chemical and biological processes, and finally the challenges and perspectives of multiscale CFD are discussed.

2. Eulerian CFD Models

The main CFD equations are derived from the first-principles laws for mass, momentum, and energy conservations in an infinitesimal volume of space [3], which are called the governing equation. The Reynolds averaged Navier–Stokes (RANS) fluid governing equations are often used in Eulerian CFD studies. The CFD models are described for single and multiple phases with the gas, liquid, and solid phases. The standard k-ε turbulence equation is presented as an example of the RANS-based turbulence equations.

2.1. Single-Phase Eulerian CFD Model

A gas or liquid phase as a continuum is modeled by a continuity equation, a Navier–Stokes (NS) momentum equation, and an energy equation [24,48]. Table 1 shows the Eulerian CFD model for an incompressible fluid. The density (ρ) is constant, and the velocity vector ($\vec{u}$) in 2D or 3D is solved from the NS equation. The stress tensor ($\bar{\bar{\tau}}$) is defined with the molecular (μ) and turbulent (μ_t) viscosities for the turbulent flow. The turbulent kinetic energy (k) is described in Section 2.5, and $\vec{g}$ is the acceleration of gravity. The temperature (T) is obtained from the energy equation, Equation (T1-3), including the convective energy (the first term in the right-hand side), the diffusive energy with conduction and viscous dissipation (the second term in the right-had side), and the heat sources (S_h) such as reaction heat and radiation. The c_p and λ are the heat capacity and effective conductivity, respectively.

Table 1. Single-phase Eulerian computational fluid dynamics (CFD) model [24,48].

Continuity equation:	$\rho\vec{\nabla}\cdot\vec{u} = 0$	(T1-1)
Momentum equation:	$\rho\frac{\partial}{\partial t}\vec{u} = -\rho\vec{\nabla}\cdot\left(\vec{u}\,\vec{u}\right) - \vec{\nabla}P + \vec{\nabla}\cdot\overline{\overline{\tau}} + \rho\vec{g}$ where $\overline{\overline{\tau}} = (\mu + \mu_t)\left[\left(\vec{\nabla}\vec{u} + \vec{\nabla}\vec{u}^T\right) - \frac{2}{3}\left(\vec{\nabla}\cdot\vec{u}\right)\overline{\overline{I}}\right] - \frac{2}{3}\rho k\overline{\overline{I}}$	(T1-2)
Energy equation:	$\rho c_p\frac{\partial T}{\partial t} = -\vec{\nabla}\cdot\vec{u}\left(\rho c_p T + P\right) + \vec{\nabla}\cdot\left(\lambda\vec{\nabla}T + \overline{\overline{\tau}}\cdot\vec{u}\right) \pm S_h$	(T1-3)

The equation of state (EOS) is needed to connect the state variables of pressure (P), volume (V), and temperature (T). As a result that Equation (T1-2) represents three equations for four unknowns (u_x, u_y, u_z, and P) with Equation (T1-1) as a constraint on the velocity, the semi-implicit method for a pressure-linked equation (SIMPLE) or the pressure implicit method with splitting the operators (PISO) is often used to link the velocity and pressure [3,49].

2.2. Gas-Liquid Eulerian CFD Model

Gas-liquid phases are modeled by the Eulerian approach assuming that the two phases flow as non-interpenetrating or interpenetrating continua. The Eulerian model assuming non-interpenetrating continua is often called the volume of fluid (VOF) method, which is a surface-tracking technique for immiscible fluids (hereafter VOF-CFD). The Eulerian model assuming interpenetrating continua is the Eulerian multiphase approach solving the continuity and momentum equations at each phase (hereafter EM-CFD).

The VOF-CFD model in an incompressible gas-liquid phase [33,50] is presented in Table 2. The VOF-CFD model relies on the assumption that the two phases do not interpenetrate [33], as mentioned earlier. The gas and liquid phases are considered as the primary and secondary phases, respectively. The VOF-CFD typically solves a single set of continuity, momentum, and energy (E) equations weighted by the phase fraction (α) [51]. The volume fraction of the secondary phase (α_L) is solved by Equation (T2-2), which is a continuity equation for the liquid phase where the interface between the phases is captured.

Table 2. Volume of fluid (VOF)-CFD model for gas-liquid phase [33,50].

Continuity equation: $\vec{\nabla}\cdot\vec{u} = 0$	(T2-1)
Volume fraction equation: $\frac{\partial\alpha_L}{\partial t} = -\vec{\nabla}\cdot\left(\alpha_L\vec{u}\right) - \vec{\nabla}\cdot\vec{u}_c\alpha_L(1 - \alpha_L)$ where the sharpening velocity $(\vec{u}_c)$ is $\vec{u}_c = C\lfloor\vec{u}\rfloor\frac{\vec{\nabla}\alpha_L}{\lfloor\vec{\nabla}\alpha_L\rfloor}$ and $\alpha_L + \alpha_G = 1$	(T2-2)
Momentum equation: $\rho\frac{\partial}{\partial t}\vec{u} = -\rho\vec{\nabla}\cdot\left(\vec{u}\,\vec{u}\right) - \vec{\nabla}P + (\mu + \mu_t)\vec{\nabla}\cdot\left[\left(\vec{\nabla}\vec{u} + \vec{\nabla}\vec{u}^T\right) - \frac{2}{3}\left(\vec{\nabla}\cdot\vec{u}\right)\overline{\overline{I}}\right] - \frac{2}{3}\rho k\overline{\overline{I}} + \rho\vec{g} + \vec{F}_{surf}$	(T2-3)
where the surface tension force is $\vec{F}_{surf} = \vec{\nabla}\cdot\overline{\overline{\tau}}_{surf}$	(T2-4)
with the surface stress tensor of $\overline{\overline{\tau}}_{surf} = \sigma\left(\left\lvert\vec{\nabla}\alpha_L\right\rvert\overline{\overline{I}} - \frac{\vec{\nabla}\alpha_L\cdot\left(\vec{\nabla}\alpha_L\right)^T}{\left\lvert\vec{\nabla}\alpha_L\right\rvert}\right)$	(T2-5)
Energy equation: $\rho\frac{\partial E}{\partial t} = -\rho\vec{\nabla}\cdot\vec{u}E + \vec{\nabla}\cdot\left(\lambda\vec{\nabla}T + \overline{\overline{\tau}}\cdot\vec{u}\right) \pm S_h$	(T2-6)
Single properties (density): $\rho = \alpha_G\rho_G + \alpha_L\rho_L$	(T2-7)
(viscosity): $\mu = \alpha_G\mu_G + \alpha_L\mu_L$	(T2-8)
(conductivity): $\lambda = \alpha_G\lambda_G + \alpha_L\lambda_L$	(T2-9)
(energy): $E = (\alpha_G\rho_G E_G + \alpha_L\rho_L E_L)/\rho$	(T2-10)

The surface tension force (F_{surf}) can be defined as a continuum surface stress (CSS) [50] tensor ($\overline{\overline{\tau}}_{surf}$) or a continuum surface force (CSF) [33]. The CSS method represents the surface tension force in a conservative manner and does not require an explicit calculation of the surface curvature. For turbulent flows, the turbulence equation is added to the VOF-CFD model (see Section 2.5).

Table 3 lists the EM-CFD model for the gas-liquid phase. The governing equation includes the continuity equation for the total mass conservation, the momentum equation with drag and non-drag forces, and the energy equation for each phase. The gas phase is considered as the secondary phase (or dispersed phase). The interfacial forces ($\vec{F}_{GL}$) between the gas and liquid phases in Equation (T3-3) can be formulated by the linear sum of the drag force ($\vec{F}_D$), the lift force ($\vec{F}_L$) emerging from the interaction between the gas (or bubble) and vorticity in the liquid [52], the wall lubrication force ($\vec{F}_W$) pushing bubbles away from the wall, and the turbulent dispersion force ($\vec{F}_T$) accounting for the turbulent momentum transfer between turbulent eddies and bubbles [53]. $\vec{F}_D$ is proportional to the difference between the gas and liquid velocities in Equation (T3-4). The gas-liquid momentum exchange coefficient (K_{GL}) contains the drag coefficient (C_D) [53]. Q_{GL} in Equations (T3-8) and (T3-9) is the heat transfer between the gas and liquid phases.

Table 3. Eulerian multiphase (EM)-CFD model for gas-liquid phase [53,54].

Continuity equation:	$\rho_G \frac{\partial}{\partial t}(\alpha_G) = -\rho_G \vec{\nabla}\cdot(\alpha_G \vec{u}_G)$ $\alpha_L = 1 - \alpha_G$	(T3-1)
	$\rho_G \frac{\partial}{\partial t}(\alpha_G \vec{u}_G) = -\rho_G \vec{\nabla}\cdot(\alpha_G \vec{u}_G \vec{u}_G) - \alpha_G \vec{\nabla}P + \vec{\nabla}\cdot\bar{\bar{\tau}}_G + \rho_G \alpha_G \vec{g} - \vec{F}_{GL}$ $\rho_L \frac{\partial}{\partial t}(\alpha_L \vec{u}_L) = -\rho_L \vec{\nabla}\cdot(\alpha_L \vec{u}_L \vec{u}_L) - \alpha_L \vec{\nabla}P + \vec{\nabla}\cdot\bar{\bar{\tau}}_L + \rho_L \alpha_L \vec{g} + \vec{F}_{GL}$ where $\bar{\bar{\tau}}_G = \alpha_G \mu_G\left(\vec{\nabla}\vec{u}_G + \vec{\nabla}\vec{u}_G^T - \frac{2}{3}\left(\vec{\nabla}\cdot\vec{u}_G\right)\bar{\bar{I}}\right) - \frac{2}{3}\rho_G k\bar{\bar{I}},$ $\bar{\bar{\tau}}_L = \alpha_L \mu_L\left(\vec{\nabla}\vec{u}_L + \vec{\nabla}\vec{u}_L^T - \frac{2}{3}\left(\vec{\nabla}\cdot\vec{u}_L\right)\bar{\bar{I}}\right) - \frac{2}{3}\rho_L k\bar{\bar{I}},$	(T3-2)
Momentum equation:	$\vec{F}_{GL} = \vec{F}_D + \vec{F}_L + \vec{F}_W + \vec{F}_T,$	(T3-3)
	$\vec{F}_D = K_{GL}(\vec{u}_G - \vec{u}_L), \quad K_{GL} = \frac{3}{4}\frac{C_D}{d_b}\rho_L \alpha_G \alpha_L\left|\vec{u}_G - \vec{u}_L\right|,$	(T3-4)
	$\vec{F}_L = -C_L \rho_L \alpha_G\left(\vec{u}_L - \vec{u}_G\right)\times\left(\vec{\nabla}\times\vec{u}_L\right),$	(T3-5)
	$\vec{F}_W = C_W \rho_L \alpha_G\left|\left(\vec{u}_L - \vec{u}_G\right)_{wall}\right|^2 \vec{n}_W,$	(T3-6)
	and $\vec{F}_T = C_T K_{GL}\frac{\mu_{t,L}}{\sigma_{GL}\rho_L}\left(\frac{\vec{\nabla}\alpha_G}{\alpha_G} - \frac{\vec{\nabla}\alpha_L}{\alpha_L}\right)$	(T3-7)
Energy equation:	$\rho_G \frac{\partial \alpha_G E_G}{\partial t} = -\rho_G \vec{\nabla}\cdot\vec{u}_G \alpha_G E_G + \vec{\nabla}\cdot\left(\alpha_G \lambda_G \vec{\nabla}T_G + \bar{\bar{\tau}}_G\cdot\vec{u}_G\right) - Q_{GL} \pm S_{h,G}$	(T3-8)
	$\rho_L \frac{\partial \alpha_L E_L}{\partial t} = -\rho_L \vec{\nabla}\cdot\vec{u}_L \alpha_L E_L + \vec{\nabla}\cdot\left(\alpha_L \lambda_L \vec{\nabla}T_L + \bar{\bar{\tau}}_L\cdot\vec{u}_L\right) + Q_{GL} \pm S_{h,L}$	(T3-9)

The mass, momentum, and heat transfers between the two phases play a key role in multiphase CFD simulations. The interfacial transfer terms require correct assessment using analytical models and empirical correlations.

2.3. Gas-Solid Eulerian CFD Model

The gas-solid CFD model is often expressed by Eulerian CFD with the kinetic theory of granular flow (KTGF) [55,56]. The KTGF-CFD model in Table 4 is the most practical choice for industrial-scale simulation of particle-fluid systems [17]. Equations (T4-1)–(T4-12) are the continuity and momentum conservation equations for the gas and solid phases. Equations (T4-17)–(T4-18) are the energy equations for each phase [57]. Here, α is the solid volume fraction, u and v are the velocities of the gas and solid phases, respectively, ρ_G and ρ_S are the densities of the gas and solid, respectively, p is the pressure shared by all phases, τ is the stress tensor associated with the two phases, and $\vec{F}_{D,GS}$ is the momentum exchange between the phases, which is defined as the interaction force per unit volume of the bed [55,58].

Table 4. Kinetic theory of granular flow (KTGF)-CFD model for gas-solid phases [55,58].

Continuity equation (gas phase): $\frac{\partial(1-\alpha)}{\partial t} + \vec{\nabla}\cdot\left((1-\alpha)\vec{u}\right) = 0$

(solid phase): $\frac{\partial\alpha}{\partial t} + \vec{\nabla}\cdot\left(\alpha\vec{v}\right) = 0$ — (T4-1)

Momentum equation (gas phase): $\rho_G\frac{\partial((1-\alpha)\vec{u})}{\partial t} = -\rho_G\vec{\nabla}\cdot\left((1-\alpha)\vec{u}\vec{u}\right) - (1-\alpha)\vec{\nabla}p + \vec{\nabla}\cdot\left(\bar{\bar{\tau}}_G\right) + \rho_G(1-\alpha)\vec{g} - \vec{F}_{D,GS}$

(solid phase): $\rho_S\frac{\partial(\alpha\vec{v})}{\partial t} = -\rho_S\vec{\nabla}\cdot\left(\alpha\vec{v}\vec{v}\right) - \alpha\vec{\nabla}p - \vec{\nabla}p_S + \vec{\nabla}\cdot\left(\bar{\bar{\tau}}_S\right) + \rho_S\alpha\vec{g} + \vec{F}_{D,GS}$ — (T4-2)

where the solid-gas drag [59] is $\vec{F}_{D,GS} = K_{GS}\left(\vec{u}-\vec{v}\right)$, $K_{GS} = \frac{3}{4}C_D\frac{\rho_G(1-\alpha)\alpha|\vec{u}-\vec{v}|}{d}(1-\alpha)^{-2.65}$

when $1-\alpha > 0.8$ (dilute), with $C_D = \frac{24}{(1-\alpha)Re_s}\left[1 + 0.15((1-\alpha)Re_s)^{0.687}\right]$, — (T4-3)

$K_{GS} = 150\frac{\alpha^2\mu_G}{(1-\alpha)d^2} + 1.75\frac{\alpha\rho_G|\vec{u}-\vec{v}|}{d}$ when $1-\alpha \leq 0.8$ (dense),

the gas-phase stress tensor is $\bar{\bar{\tau}}_G = (1-\alpha)\mu_G\left[\vec{\nabla}\vec{u} + \vec{\nabla}\vec{u}^T - \frac{2}{3}\left(\vec{\nabla}\cdot\vec{u}\right)\bar{\bar{I}}\right] - \frac{2}{3}\rho_G k\bar{\bar{I}}$, — (T4-4)

the solid-phase stress tensor is $\bar{\bar{\tau}}_S = \left[-p_S + \alpha\lambda_S\vec{\nabla}\cdot\vec{v}\right]\bar{\bar{I}} - \alpha\mu_S\left[\vec{\nabla}\vec{v} + \vec{\nabla}\vec{v}^T - \frac{2}{3}\left(\vec{\nabla}\cdot\vec{v}\right)\bar{\bar{I}}\right] - \frac{2}{3}\rho_G k\bar{\bar{I}}$, — (T4-5)

the solid-phase pressure is $p_S = \alpha\rho_S\left[1 + 2(1+e)\alpha g_0\right]\Theta$, — (T4-6)

the solid-phase bulk viscosity [60] is $\lambda_S = \frac{4}{3}\alpha\rho_S d g_0(1+e)\left(\frac{\Theta}{\pi}\right)^{\frac{1}{2}}$, — (T4-7)

the solid shear viscosity is $\mu_S = \mu_{S,col} + \mu_{S,kin} + \mu_{S,fri}$, — (T4-8)

the collision viscosity of solids [61] is $\mu_{S,col} = \frac{4}{5}\alpha\rho_S d g_0(1+e)\left(\frac{\Theta}{\pi}\right)^{\frac{1}{2}}$, — (T4-9)

the kinetic viscosity of solids [59] is $\mu_{S,kin} = \frac{10\rho_S d\sqrt{\Theta\pi}}{96\alpha(1+e)g_0}\left[1 + \frac{4}{5}g_0\alpha(1+e)\right]^2\alpha$, — (T4-10)

the frictional viscosity of solids [62] is $\mu_{S,fri} = \frac{p_s\sin\phi}{2\sqrt{I_{2D}}}$; with $\phi = 30^0$, — (T4-11)

and the radial distribution function [63] is $g_0 = \left[1 - \left(\frac{\alpha}{\alpha_m}\right)^{\frac{1}{3}}\right]^{-1}$ with the maximum packing (α_m) — (T4-12)

Granular temperature (Θ) equation neglecting the convection and diffusion terms: $0 = \left(-p_S\bar{\bar{I}} + \bar{\bar{\tau}}_S\right):\vec{\nabla}\vec{v} - \gamma_S + \phi_{GS}$ — (T4-13)

where the energy generation by the solid stress tensor $(\bar{\bar{\tau}}_S)$ with a double inner product is $\left(-p_S\bar{\bar{I}} + \bar{\bar{\tau}}_S\right):\vec{\nabla}\vec{v}$, — (T4-14)

the energy collisional dissipation rate of the solid phase [60] is $\gamma_S = \frac{12(1-e^2)g_0}{d\sqrt{\pi}}\rho_S\alpha^2\Theta^{\frac{3}{2}}$, — (T4-15)

and the fluctuation kinetic energy transfer from the solid to gas phase [59] is $\phi_{GS} = -3K_{GS}\Theta$ — (T4-16)

Energy equation: $\rho_G\frac{\partial(1-\alpha)E_G}{\partial t} = -\rho_G\vec{\nabla}\cdot\vec{u}(1-\alpha)E_G + \vec{\nabla}\cdot\left((1-\alpha)\lambda_G\vec{\nabla}T_G + \bar{\bar{\tau}}_G\cdot\vec{u}\right) - Q_{GS} \pm S_{h,G}$ — (T4-17)

$\rho_S\frac{\partial\alpha E_S}{\partial t} = -\rho_S\vec{\nabla}\cdot\vec{v}\alpha E_S + \vec{\nabla}\cdot\left(\alpha\lambda_S\vec{\nabla}T_S + \bar{\bar{\tau}}_S\cdot\vec{v}\right) + Q_{GS} \pm S_{h,S}$ — (T4-18)

Assuming incompressible flow (constant ρ), no chemical reaction, and no phase change, the first term of Equation (T4-1) accounts for the rate of total mass accumulation per unit volume, and the second term is the net rate of convection mass flux. $\bar{\bar{\tau}}_G$ and $\bar{\bar{\tau}}_S$ are expressed in Equations (T4-4) and (T4-5), respectively, for a Newtonian fluid with constant viscosity (μ).

The Gidaspow drag model (K_{GS}) [59] in Equation (T4-3), which is a combination of the Ergun and Wen–Yu drag models, is applicable to both dilute and dense solids flows [21]. The drag model was validated by a number of studies for bubbling fluidized bed with acceptable accuracy [55]. The granular temperature (Θ) in Equation (T4-6) for the solid pressure (p_S) is defined with the energy generation by the solid stress in Equation (T4-14), the dissipation rate (γ_S) of collisional energy in Equation (T4-15), and the random fluctuation kinetic energy transfer (ϕ_{GS}) from the solid to gas phases in Equation (T4-16) [55,56].

The mesoscale structure of clustered particles has a significant effect on the flow and transport behavior of particle-fluid systems [17]. The drag coefficient $(C_D$ in Equation (T4-3)) measured in gas-solid fluidization may be orders of magnitude lower than that of uniform suspension because of the presence of bubbles and particle clusters [17]. To elucidate the complex behavior of clustered particles, energy minimization multiscale (EMMS) methods have been proposed, where the energy consumption for the suspension and transport of particles is minimized under stability constraints [64].

Using the operating conditions (superficial gas velocity and solid circulation flux) and the physical properties of the gas (density and viscosity) and the solid (diameter and density) as input parameters, the drag coefficient of the EMMS method is calculated in sequential or concurrent strategy coupling with the KTGF-CFD model in the macroscopic scale [17]. The table-looking method as a sequential coupling strategy can be used for the EMMS method in a computationally efficient way (refer to Section 3.5 for details) [58].

2.4. Three-Phase Eulerian CFD Model

In the three-phase Eulerian CFD model, three phases are treated as fully interpenetrating continuous phases. The solid pressure (p_S) and viscosity (μ_S) are calculated by the KTGF model. All the phases have the same conservation equations, where additional closures (or constitutive equations) are required for interface interactions such as gas-liquid ($\vec{F}_{D,GL}$), solid-liquid ($\vec{F}_{D,LS}$), and gas-solid ($\vec{F}_{D,GS}$) drags [23], and gas-liquid (Q_{GL}), solid-liquid (Q_{LS}), and gas-solid (Q_{GS}) heat exchanges [57]. The drag forces ($\vec{F}_D$) caused by the slip velocity between phases are predominant in the interphase momentum exchange forces, therefore Table 5 shows only the drag forces in the momentum equation.

Table 5. EM-KTGF-CFD model for gas-liquid-solid phases [65].

$$\text{Continuity equation (gas phase): } \rho_G\frac{\partial \alpha_G}{\partial t} = -\rho_G\vec{\nabla}\cdot\left(\alpha_G\vec{u}_G\right)$$
$$\text{(liquid phase): } \rho_L\frac{\partial \alpha_L}{\partial t} = -\rho_L\vec{\nabla}\cdot\left(\alpha_L\vec{u}_L\right) \tag{T5-1}$$
$$\text{(solid phase): } \rho_S\frac{\partial \alpha_S}{\partial t} = -\rho_S\vec{\nabla}\cdot\left(\alpha_S\vec{u}_S\right)$$

$$\text{Momentum equation (gas phase): } \rho_G\frac{\partial\left(\alpha_G\vec{u}_G\right)}{\partial t} = -\rho_G\vec{\nabla}\cdot\left(\alpha_G\vec{u}_G\vec{u}_G\right) - \alpha_G\vec{\nabla}p + \vec{\nabla}\cdot\bar{\bar{\tau}}_G + \rho_G\alpha_G\vec{g} - \vec{F}_{D,GL} - \vec{F}_{D,GS}$$
$$\text{(liquid phase): } \rho_L\frac{\partial\left(\alpha_L\vec{u}_L\right)}{\partial t} = -\rho_L\vec{\nabla}\cdot\left(\alpha_L\vec{u}_L\vec{u}_L\right) - \alpha_L\vec{\nabla}P + \vec{\nabla}\cdot\bar{\bar{\tau}}_L + \rho_L\alpha_L\vec{g} + \vec{F}_{D,GL} - \vec{F}_{D,LS} \tag{T5-2}$$
$$\text{(solid phase): } \rho_S\frac{\partial\left(\alpha_S\vec{u}_S\right)}{\partial t} = -\rho_S\vec{\nabla}\cdot\left(\alpha_S\vec{u}_S\vec{u}_S\right) - \alpha_S\vec{\nabla}p - \vec{\nabla}p_S + \vec{\nabla}\cdot\bar{\bar{\tau}}_S + \rho_S\alpha_S\vec{g} + \vec{F}_{D,LS} + \vec{F}_{D,GS}$$

$$\text{Interface interactions (G-L): } \vec{F}_{D,GL} = K_{GL}\left(\vec{u}_G - \vec{u}_L\right),\ K_{GL} = \frac{3}{4}\frac{C_{D,GL}}{d_b}\rho_L\alpha_G\alpha_L\left|\vec{u}_G - \vec{u}_L\right|$$
$$\text{(S-L): } \vec{F}_{D,SL} = K_{SL}\left(\vec{u}_S - \vec{u}_L\right),\ K_{SL} = \frac{3}{4}C_{D,SL}\frac{\rho_L\alpha_S\alpha_L\left|\vec{u}_S-\vec{u}_L\right|}{d_s}\alpha_L^{-2.65}\ for\ \alpha_L > 0.8,$$
$$and\ K_{SL} = 150\frac{\alpha_S(1-\alpha_L)\mu_L}{\alpha_L d_s^2} + 1.75\frac{\alpha_S\rho_L\left|\vec{u}_S-\vec{u}_L\right|}{d_s}\ for\ \alpha_L \leq 0.8 \tag{T5-3}$$
$$\text{(G-S): } \vec{F}_{D,GS} = K_{GS}\left(\vec{u}_G - \vec{u}_S\right),\ K_{GS} = \frac{3}{4}\frac{C_{D,GS}}{d_s}\rho_g\alpha_G\alpha_S\left|\vec{u}_G - \vec{u}_L\right|$$

$$\text{Energy equation (gas phase): } \rho_G\frac{\partial\alpha_G E_G}{\partial t} = -\rho_G\vec{\nabla}\cdot\vec{u}_G\alpha_G E_G + \vec{\nabla}\cdot\left(\alpha_G\lambda_G\vec{\nabla}T_G + \bar{\bar{\tau}}_G\cdot\vec{u}_G\right) - Q_{GL} - Q_{GS} \pm S_{h,G}$$
$$\text{(liquid phase): } \rho_L\frac{\partial\alpha_L E_L}{\partial t} = -\rho_L\vec{\nabla}\cdot\vec{u}_L\alpha_L E_L + \vec{\nabla}\cdot\left(\alpha_L\lambda_L\vec{\nabla}T_L + \bar{\bar{\tau}}_L\cdot\vec{u}_L\right) + Q_{GL} - Q_{LS} \pm S_{h,L} \tag{T5-4}$$
$$\text{(solid phase): } \rho_S\frac{\partial\alpha_S E_S}{\partial t} = -\rho_S\vec{\nabla}\cdot\vec{u}_S\alpha_S E_s + \vec{\nabla}\cdot\left(\alpha_S\lambda_s\vec{\nabla}T_s + \bar{\bar{\tau}}_s\cdot\vec{u}_S\right) + Q_{LS} + Q_{GS} \pm S_{h,S}$$

Pan et al. (2016) [23] summarized the drag coefficient ($C_{D,GL}$) between the gas and liquid phases such as in the Grace, Tomyama, Schiller–Naumann, and Ishii–Zuber models. The gas-liquid drag force influences the gas holdup. Hamidipour et al. (2012) [65] used the Gidaspow drag model [59] to describe the solid-liquid interaction force ($\vec{F}_{D,SL}$), which has been applied to gas-solid systems. The gas-solid drag force ($\vec{F}_{D,GS}$) similar to that between the gas and liquid phases is used in Equation (T5-3). In addition to the drag models, the granular temperature (Θ) from the KTGF is calculated (see Equation (T4-13)–Equation (T4-16)) for the solid pressure (p_S) and viscosity (μ_S) [65].

2.5. Turbulence Equations

The dynamics of fluid flow are described by the NS momentum equations [31]. As the Reynolds number increases, the flow reaches a state of motion characterized by strong and unsteady random fluctuations of the velocity and pressure fields, which is referred to as the turbulent regime. Turbulent flow can be modeled by the direct numerical simulation (DNS) of the NS equations, which is computationally prohibitive but accurate, RANS based on empirical models, and large eddy simulation (LES) as a compromise between DNS and RANS [31].

The instantaneous velocity and pressure are decomposed into the sum of the mean and fluctuation components in the RANS-based equation. Substituting the decomposition into the NS equations and taking the ensemble-average leads to the RANS equations. RANS modeling includes uncertainties due to information loss in the Reynolds-averaging process [31]. Here, the Reynolds stress tensor ($\bar{\bar{\tau}}$) including the turbulent viscosity (μ_t) is solved with constitutive equations describing the turbulent kinetic energy (k) and its dissipation rate (ε). The RANS equation for the turbulent momentum equation is expressed as

$$\rho\vec{\nabla}\cdot\left(\vec{u}\,\vec{u}\right) = -\vec{\nabla}P + \vec{\nabla}\cdot\bar{\bar{\tau}} + \rho\vec{g} + \vec{F} \tag{1}$$

where the Reynolds stress tensor ($\bar{\bar{\tau}}$) is defined as

$$\bar{\bar{\tau}} = (\mu + \mu_t)\left(\vec{\nabla}\vec{u} + \vec{\nabla}\vec{u}^T - \frac{2}{3}\left(\vec{\nabla}\cdot\vec{u}\right)\bar{\bar{I}}\right) - \frac{2}{3}\rho k\bar{\bar{I}} \tag{2}$$

with $\mu_t = \rho C_\mu \frac{k^2}{\varepsilon}$ and $\bar{\bar{I}} = \delta_{ij}$, $\delta_{ij} = 1 \; for \; i = j \; and \; \delta_{ij} = 0 \; for \; i \neq j$.

The standard k-ε turbulence model [50,66,67] is expressed for incompressible fluid as follows:

$$\rho\frac{\partial k}{\partial t} + \rho\vec{\nabla}\cdot\left(k\vec{u}\right) = \vec{\nabla}\cdot\left[\left(\mu + \frac{\mu_t}{\sigma_k}\right)\vec{\nabla}k\right] + G_k + G_b - \rho\varepsilon \tag{3}$$

$$\rho\frac{\partial\varepsilon}{\partial t} + \rho\vec{\nabla}\cdot\left(\varepsilon\vec{u}\right) = \vec{\nabla}\cdot\left[\left(\mu + \frac{\mu_t}{\sigma_\varepsilon}\right)\vec{\nabla}\varepsilon\right] + C_{\varepsilon1}\frac{\varepsilon}{k}G_k - C_{\varepsilon2}\rho\frac{\varepsilon^2}{k} \tag{4}$$

where G_k and G_b represent the generation of the turbulent kinetic energy (k) caused by the mean velocity gradients and the buoyancy, respectively.

The k-ε model involves coefficients C_μ, $C_{\varepsilon1}$, $C_{\varepsilon2}$, σ_k, and σ_ε. The nature of these coefficients leads to ambiguity regarding their values. However, these coefficients have been calibrated empirically to reproduce the results of a few flows [31]. Xiao and Cinnella (2019) presented the uncertainty of these parameters [31]. The RANS-based turbulence model [26] includes the standard k-ε [67], modified k-ε, realizable k-ε for complex flows [22,24,55,68], renormalization group (RNG) k-ε, and shear stress transport (SST) k-ω equations [53]. The RNG k–ε model is more accurate and reliable for a wider range of flows than the standard k–ε model, which is mostly used in turbulence modeling of duct channel flows with heat transfer [40]. The realizable k-ε turbulence model satisfies the mathematical constraints on the Reynolds stresses, which is consistent with the physics of turbulence flow. The realizable k-ε model has exhibited substantial improvements over the other k-ε type turbulence models when the flow features include strong streamline curvature, vortex, and rotation [68]. The SST k-ω model is widely used for laminar and turbulent mixed flows [53,69].

For turbulent two-phase flows in microscale CFD, LES may offer a good compromise between RANS-based closures and DNS due to suitable subgrid closures [32], as mentioned earlier. However, wide application of LES is still limited because of the large mesh requirement and high computational cost [39]. The DNS and LES may be impractical as a general-purpose design tool for most industrial flow conditions [39].

3. Applications of Eulerian CFD

CFD makes it possible to understand complex hydrodynamic behaviors with heat and mass transfer and chemical reactions. CFD has numerous applications in the field of gas distributors [24], gas storage tanks [26], coal and biomass gasifiers [55,70], fuel cells [38,71], absorbers with random packing [72–74] and structured-packing [54,75], agitated bioreactors [39,76], offshore oil separators [77], gas-liquid bubble columns [32,53,78–80], and gas-solid fluidized beds [23,43,55,56,58,65].

3.1. Single-Phase CFD Simulations

Gas-phase CFD with various turbulence equations has been used to find the optimal geometries of equipment such as a vortex tube producing a temperature separation effect [81], a syngas conversion reactor [42], a duct with rib-roughened walls [40], an industrial-scale gas distributor with injectors [24], and a sleeve-type steam methane reformer for H_2 production [22]. For oxy-fuel combustion in power plants, a gas-phase Eulerian CFD model was incorporated with combustion reactions (pyrolysis, gasification, char reactions, and gas-phase reactions), turbulence, convective and radiative heat transfers, and convective mass transfer [41].

Using a liquid-phase 3D CFD model with heat transfer, the effect of various hollow fiber geometries on hydrodynamics through the flow channels was investigated to enhance the heat transfer coefficient (HTC) between the feed and the permeate [82]. Liquid-phase CFD under a laminar flow condition was used to evaluate the HTC of nanoparticle-added ionic liquids that are non-flammable and non-volatile at ambient conditions and recyclable [83]. The effect of the Reynolds number and nanoparticle concentration on the HTC was investigated using the Eulerian CFD model with the thermos-physical properties of ionic nanofluids such as thermal conductivity, heat capacity, density, and viscosity [83].

3.1.1. Gas Distributors

CFD is considered a pivotal technology in the design and performance evaluation of chemical processes [84]. Appropriate performance indexes should be obtained from CFD results for both the process design and optimization. Pham et al. (2018) [24] compared several gas distributors used in an industrial-scale CO_2 amine absorber using a gas-phase steady-state 3D CFD with the realizable k-ε turbulence equation. Three performance indexes, including the pressure drop, dead-area ratio, and coefficient of gas distribution, were used to describe the quality of gas distribution [24]. Figure 2 shows a Schoepentoeter gas distributor as a gas injection device, the velocity vector around the Schoepentoeter, and the radial velocity contour above the chimney tray. The velocity contour obtained from the CFD simulation (Figure 2c) is able to quantify the uniformity of the gas distribution.

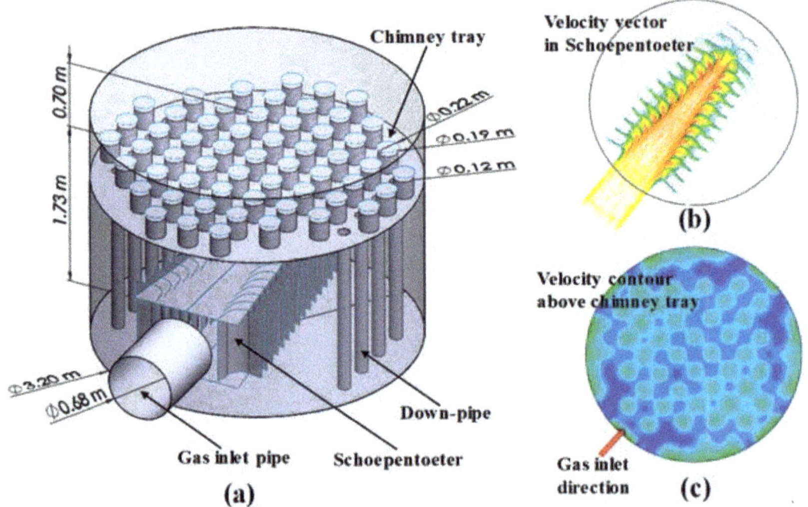

Figure 2. Schoepentoeter gas distributor used in an amine absorber. (**a**) Gas distributor with Schoepentoeter; (**b**) gas velocity vector around Schoepentoeter; and (**c**) radial velocity contour above chimney tray (modified from Pham et al. (2018) [24]).

3.1.2. Gas-Filling Storage Processes

The maximum temperature is regulated throughout H_2 refueling to preserve the integrity of the tanks [26]. CFD simulations of H_2 tanks with experimental validation were performed in the range of $50 \leq P \leq 800$ bar [26]. CFD made it possible to investigate the effect of feed temperature, density, pressure, and flow speed on the gas and tank wall temperatures in space and time [26]. A gas-phase CFD model with internal and external heat transfers, which was validated with experimental data, was used to identify the temperature profile of the H_2 tank.

3.1.3. Fixed Bed with Random Packing

Ahmadi and Sefidvash (2018) [72] presented a gas-phase steady-state 3D CFD with the standard k-ε turbulence equation to predict the pressure drop in a fixed bed with random packing. A supercritical steam flow with high Reynolds numbers was simulated in a column randomly packed with 800 fixed particles of 3 cm in diameter. After validation of the CFD model against experimental data, the CFD model was used to develop a CFD-based empirical equation for the pressure drop at high Reynolds numbers in random packing fixed beds, which is useful in industry [72]. Unlike porous media CFD simulations for fixed beds [75,85], this CFD study considered the random structure of 800 spherical particles using 15 million computational cells.

Wang et al. (2020) [73] reported a gas-phase steady-state 3D CFD with the realizable k-ε turbulence equation to predict the pressure drop in a rotating fixed bed with random packing. N_2 gas flow was simulated in a rotating column randomly packed by 470 fixed particles of 5 mm in diameter. Approximately 10 million computational cells were used to build a fixed bed randomly packed with the spherical particles. The CFD model was first validated with experimental data. Subsequently, the effects of rotating speed and gas velocity on the pressure drop were examined using the validated CFD model. A semi-empirical correlation based on the CFD simulation was proposed to predict the dry pressure drop [73], which is of practical use in industry.

3.2. Gas-Liquid Phase CFD Simulations

CFD simulations in the gas-liquid phase have been applied to chemical and biological processes such as fuel cells [38,86], microchannel reactors [87], WWTPs [39], photobioreactors [76], absorbers with structured-packing [54,75] and random packing [74], and bubble columns [53,78–80]. VOF-CFD was used to simulate relatively small physical domains, whereas EM-CFD was used to investigate the hydrodynamic behaviors of an entire process.

Using VOF-CFD models, the effects of operating conditions (temperature and gas flow rate), material properties, and gas channel geometries on the hydrodynamics of PEM fuel cells (cathode side) were reported by Ferreira et al. (2015) [38]. Multiple length scales from a few micrometers to meters were also present in the PEM fuel cell. Coupling of the VOF model with electrochemical reactions and heat transfer is needed to better understand the hydrodynamics of PEM fuel cells [38]. Furthermore, the coupling of gas-liquid flows in both anode/cathode channels and porous electrodes is vital in current CFD models, which leads to multiscale simulations [86]. Zhang and Jiao (2018) [86] presented the necessity of a 3D CFD simulation at cell and stack levels to accurately simulate water and thermal management in PEM fuel cells.

Li et al. (2018) [87] used a 3D VOF-CFD model with CSF to investigate the pressure drop, liquid holdup, and gas-liquid interfacial area in a microporous tube-in-tube microchannel reactor. The pressure drop obtained from the CFD was validated against that of an experiment. The effects of the microchannel width, micropore diameter, and liquid flow rate on the pressure drop and interfacial area were examined using the CFD model [87]. The CFD study addressed the potential for optimization of operating conditions and design of microchannel reactors.

CFD tools have been used as a design tool in the development of new aeration devices and optimization of their geometry in WWTPs [39]. The gas-liquid EM-CFD model was applied to activated sludge tanks of WWTP to investigate the gas holdup, flow velocity, and O_2 concentration [39]. The fluid flow in such bioreactors depended on the tank geometry, physical properties (density, viscosity, and solubility), and operating conditions (inlet flow rate, inlet concentration, and temperature) [39]. A CFD model combining hydrodynamics, mass transfer, and biochemical reaction kinetics will be one of the major challenges in WWTP [39]. Collision, coalescence, breakage, and deformation of bubbles promote diversity in the shapes and sizes of bubbles in aeration tanks with a bubbly flow regime. The population balance equation (PBE) for bubbles may be helpful in accounting for the bubble-size distribution (BSD) and its dynamics [39].

Soman and Shastri (2015) [76] presented a gas-liquid 3D Eulerian CFD with radiation in a large-scale photobioreactor for microalgae cultivation. The dimensions of a new photobioreactor (width, height, and clearances) were optimized using the CFD model validated against experimental data. The bubble size was set to 5 mm in the CFD. Assuming that the microalgae was distributed uniformly and was well diluted, the solid phase was ignored [76].

3.2.1. Gas-Liquid Flow in Structured Packing

Kim et al. (2017) [75] investigated the effect of the center of gravity (CoG) position on the CO_2 removal efficiency for an amine absorber with structured packing subject to pitching motion. A gas-liquid transient Eulerian 3D CFD model was used. The sliding mesh method observing at the outside of the amine absorber was implemented for the pitching motion [75]. The structured packing was assumed as a porous medium, ignoring complex flow channels of the packing. However, porous resistance and liquid dispersion forces were added into the NS momentum equation in the porous medium zone to mimic key features of the packing [54].

Figure 3a shows the layout of the three CoGs used for comparison in the amine absorber (0.125 m in diameter and 4.325 m in height). The equipment motion is the least severe when the equipment is located as close as possible to the CoG. When the column was inclined to the left, the liquid holdup of CoG3 in the middle of the absorber (h = 2.1 m) was strongly skewed to the left wall side (see Figure 3b). The CFD study successfully identified the effect of the installed location on the hydrodynamics of the process under ship motion. The hydrodynamics of an air-water-oil separator under three angular ship motions (pitch, roll, and yaw) were also presented recently [77].

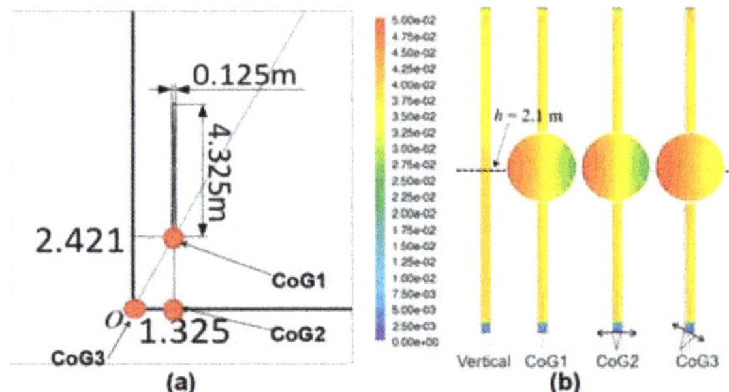

Figure 3. Amine absorber with structured-packing under off-shore operation. (**a**) Layout of amine absorber with three centers of gravity (CoG); and (**b**) liquid holdup contour under pitching motion at h = 2.1 m for three CoGs (modified from Kim et al. (2017) [75]).

3.2.2. Gas-Liquid Flow in Random Packing

Kang et al. (2019) [74] presented an unsteady-state 3D VOF-CFD that included a surface tension force (i.e., CSF) to investigate gas-liquid flow behaviors in random packing with 80 Rasching rings (inner diameter 4 mm and outer diameter 6 mm). The VOF-CFD results were compared with experimental data such as the gas-liquid interfacial area, liquid holdup, and pressure drop. The CFD study represented well the hydrodynamics of the small-scale packing column [74]. However, a VOF-CFD approach that considers the detailed geometry of packing particles may be prohibitive for a large-scale industrial column, as mentioned earlier. The microscale VOF-CFD simulation may be useful for developing correlations that are required in a macroscale CFD simulation for industrial packing columns [74].

3.2.3. Gas-Liquid Bubble Columns

Bubble columns have been widely used in chemical, pharmaceutical, petrochemical, biochemical, and metallurgical processes for gas-liquid and gas-liquid-solid contact or chemical reactions due to good heat and mass transfer [25,53,78,88]. CFD has become an important approach in understanding the hydrodynamics of multiphase flow, bubble-size distribution, and gas-liquid interfacial area, thus facilitating the design, scaling-up, or optimization of bubble columns [25]. Many researchers developed CFD models with various constitutive equations or sub-models for bubble columns.

Klein et al. (2019) presented a VOF-CFD simulation with an LES turbulence equation for liquid jet atomization and a single bubble rising in turbulent flow at the microscale [32]. For industrial-scale bubble column fermenters, Fletcher et al. (2017) [78] presented a transient 3D EM-CFD without the population balance equation (PBE) to describe bubble breakage and coalescence. The drag model was modified considering bubble swarms, and the mass transfer of O_2 from the gas to liquid phase was considered via the linear driving force model. The CFD model was validated against experiments (radial gas holdup and radial liquid velocity). In the CFD study, the use of a single bubble size was a gross simplification. The impact of bubble swarms and wakes on the drag force must be better understood. Bubble-induced turbulence [79] may improve the level of agreement between modeling and experimental results [78].

A wide range of bubble sizes and shapes exist in bubbly flows, which leads to different transport characteristics. Sharma et al. (2019) [79] proposed turbulence-induced bubble collision diffusion for two-group bubble simulations. The bubble collision force for the two groups of bubbles was added to the gas-phase momentum equation of each group [79]. The gas-liquid Eulerian CFD model without PBE was applied to a turbulent flow regime. Bubble interaction mechanisms such as bubble breakup and coalescence would be useful to better predict hydrodynamics in bubble columns.

Tran et al. (2019) [53] presented a gas-liquid Eulerian CFD model coupled with a PBE for an air-water homogeneous bubble column under high pressures. The drag coefficient and breakage kernel were modified to consider the impact of bubble swarms and pressure on the hydrodynamics of the bubble column. The gas holdup and mean bubble size obtained from the modified CFD model were compared with experimental data. Figure 4 displays the gas holdup contours in the air-water bubble column at 1–35 bar. The gas holdup increases as the pressure increases. It is clearly shown at the bottom of Figure 4 that the radial gas holdup is higher at higher pressure.

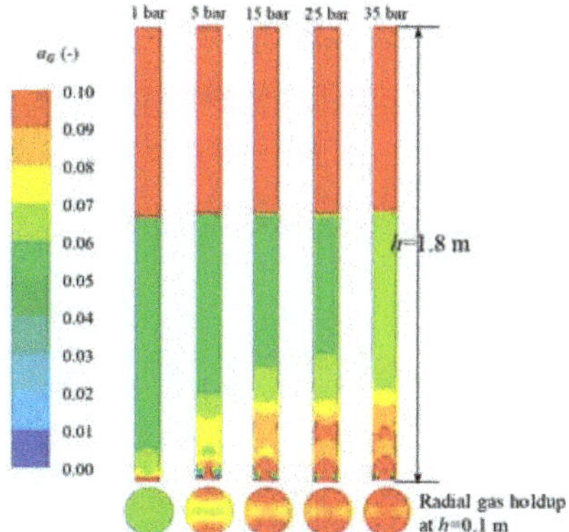

Figure 4. Gas holdup contours in air-water homogeneous bubble column at 1–35 bar (modified from Tran et al. (2019) [53]).

Yan et al. (2019) [80] proposed three drag models used for a gas-liquid transient Eulerian CFD model without PBE in an air-water bubble column under elevated pressure. The drag models covering the flow regime from homogeneous to turbulent included a pressure correction term to represent the effect of pressure on the gas holdup [80]. Gas-organic liquid bubble columns at high pressures and temperatures are common in industrial applications [89], therefore attention has been paid to the CFD simulation of those bubble columns. However, prior to CFD simulations for high-pressure and high-temperature bubble columns, reliable drag, breakage, and coalescence models have to be developed that consider microscopic bubble behaviors [90].

Through the stability condition at a mesoscale flow structure, Yang and Xiao (2017) [25] presented an approach based on the EMMS concept to calculate correction factors of the drag coefficient (C_D) and the coalescence rate. A CFD-PBE model coupled with the EMMS concept was solved for air-water bubble columns at various gas velocities and compared to experimental data such as the axial bubble size and bubble size distribution. The CFD-PBE model considering the mesoscale bubble structure showed improved prediction of the experimental data over the original CFD-PBE [25].

3.3. Gas-Solid Phase CFD Simulations

Fluidized-bed combustors (FBCs) such as bubbling and circulating FBCs have been used for power production from coal, biomass, and organic wastes owing to excellent heat and mass transfer, low pollutant emission, and high combustion efficiency [91,92]. Fluidized-bed chemical looping combustion (FB-CLC) has shown promise as a solution for high-efficiency low-cost carbon capture from fossil-fueled power plants [93]. 2D and 3D Eulerian CFD simulations with chemical reactions of the FB-CLC were performed to understand hydrodynamics coupled with chemical reactions [93].

Fluidized beds can be formulated using the two-fluid CFD model where the solid phase is considered to be a continuum and fully interpenetrating with the gas phase. A Eulerian–Eulerian CFD framework involving a granular solid and a gas phase is often used for computational investigations of fluidized beds [91]. The Eulerian CFD model includes the governing equations (the continuity equation, NS momentum equations, and energy equation) and the constitutive equations for the solid-phase pressure and viscosity provided by the KTGF [93], as listed in Table 4.

Zhou and Wang (2015) investigated mixing and segregation in circulating fluidized-bed (CFB) risers using the KTGF-CFD model with the EMMS drag model. The EMMS approach showed improved hydrodynamics over homogeneous drag forces compared to the experimental data [58]. Fotovat et al. (2017) [43] presented Eulerian and Lagrangian CFD models to investigate the electrostatic effects on the hydrodynamics of fluidized beds. The CFD models assumed constant particle charges without allowing for the dynamic nature of tribo-electrification. Thus, a multiscale approach where charge generation and dissipation mechanisms at the particle scale are coupled with the distribution of charged particles at the mesoscale was suggested for the future work to predict the effect of electrostatic charges on the overall reactor performance [43].

The computational particle fluid dynamics (CPFD) model as a kind of Eulerian–Lagrangian method has been applied to gas-solid fluidized beds [92]. The solid phase is modeled by the Liouville equation for a particle distribution function (f) in CPFD [94,95] instead of the KTGF equation in Eulerian CFD. The solid properties are calculated from the f of representative computational particles, which can increase the computational efficiency for dense solid phases [92]. Kim et al. (2020) [96] presented a CPFD coupled with the PBE, which was applied to a multiphase and multiscale crystallization. Wu et al. (2020) [92] developed a cluster-based drag model to consider the multiscale characteristics of particles in a CPFD simulation.

3.4. Gas-Liquid-Solid Phase CFD Simulations

Yang et al. (2001) [97] investigated numerically and experimentally the bubble formation in a gas-liquid-solid fluidized bed at elevated pressures. The 2D VOF-CFD successfully predicted the bubble size with respect to the pressure and solid fraction using the liquid density, viscosity, and surface tension adjusted with the pressure. The VOF-CFD was performed in a small area (8 cm width and 5 cm height) with two orifices at the bottom [97].

Hamidipour et al. (2012) [65] presented an unsteady-state 3D EM-KTGF-CFD simulation of gas-liquid-solid fluidized-beds. The CFD results obtained from various turbulence equations were compared with the experimental data such as the axial solid velocity and gas holdup. This study indicated that the constitutive equations and numerical schemes involved in the CFD model should be carefully selected to accurately capture the flow patterns.

Pan et al. (2016) reviewed CFD simulations of gas-liquid-solid flow in fluidized beds [23]. A three-phase fluidized bed is intrinsically multiscale in time and space, including the macroscale at the reactor level, the mesoscale with particle clusters and bubble swarms, and the microscale with particle and bubble size [23]. Hydrodynamic properties such as the bubble size, bubble-rising velocity, liquid velocity, and gas and solid holdups were examined according to operating conditions in three-phase fluidized beds. The authors suggested additional work required for three-phase fixed-bed CFDs: (i) new drag models for interphase momentum exchange, (ii) CFD modeling including chemical reactions and heat and mass transfers in industrial-scale reactors, and (iii) multiscale modeling on the balance between computation cost and accuracy [23].

3.5. Multiscale CFD Simulations

The critical task and major difficulty in multiscale modeling is how to exchange information at the interface of neighboring sub-domains [98]. The applications of the multiscale CFD simulation are introduced in terms of concurrent and sequential coupling strategies.

3.5.1. Concurrent Coupling Strategy

Chen et al. (2013) [98] presented a multiscale modeling framework combining a gas-phase Eulerian CFD and a D2Q5/D2Q9 LBM to predict electrochemical transport phenomena in the cathode of PEM fuel cells. At each time step, the density, velocity, and concentrations were interchanged between the CFD and LBM domains, including the gas channel, porous gas diffusion layer, and catalytic layer. The results obtained from LBM were used as the boundary conditions of the CFD domain. At each time step, the CFD and LBM simulations were repeated until the two results converged [98].

Crose et al. (2017) [13] developed a multiscale CFD framework in the plasma-enhanced chemical vapor deposition (PECVD) of thin-film solar cells. A macroscopic gas-phase transient 2D CFD model with plasma chemistry and transport phenomena was solved to calculate the pressure, velocity, temperature, and species concentration. The CFD model was concurrently coupled with a kinetic Monte Carlo scheme describing the microscopic thin-film growth. Indeed, decoupled CFD and surface interaction models were unable to capture the spatially nonuniform deposition of silicon films. The multiscale CFD was capable of predicting the codependent behavior of the macroscopic gas phase and the microscopic thin-film growth rate. Using three user-defined functions (UDFs), including gas-phase reaction rates, an electron density profile in the gas phase, and a hybrid kinetic Monte Carlo (KMC) algorithm on the wafer surface, the macro- and microscale models were coupled at each time step [13]. The multiscale computing was parallelized within a message-passing interface (MPI) architecture to reduce the computational time.

Pozzetti and Peters (2018) [99] proposed a multiscale DEM-CFD method for the simulation of three-phase flows using a dual-grid method for the bulk and fluid fine scales, which is different from the classical CFD-DEM approach. The classical CFD-DEM approach based on the volume averaging technique aims to couple CFD with a DEM solution at a single scale using cell values or locally interpolated values to evaluate the fluid-particle interactions. The concurrent coupling between CFD and DEM can lead to high computational cost. The dual-grid multiscale DEM-CFD method aimed to capture the most significant phenomena of the two different scales while reducing the computational burden as much as possible. However, the new method required a complex interpolation between the two scales [99].

A heterogeneity concept for mesoscale particle-cluster behaviors was introduced into Eulerian CFD computational cells to modify the interphase drag coefficient in the fluidized bed. Qi et al. (2007) [100] presented EMMS-based multiscale drag models, where flow variables such as the voidage (ε) and the gas and solid velocities (u_g and u_s, respectively) obtained from CFD were used as input values to the EMMS model. The drag force (F_D) considering the heterogeneity of the mesoscale was simultaneously updated for each computational cell and flow time [100]. Li et al. (2018) [87] proposed a cluster structure-dependent (CSD) drag model by introducing convective and temporal accelerations of both gas and particles in the dilute and dense phases. Although multiscale drag models showed better accuracy than common single-scale drag models, they required greater computational time.

The concurrent coupling strategy of the gas-solid Eulerian CFD with the multiscale drag models is shown in Figure 5. In each computational cell, an EMMS or CSD model is solved with the input variables (ε, u_g, and u_s) from the CFD. The drag force (F_D) computed by the EMMS or CSD model is returned to each cell.

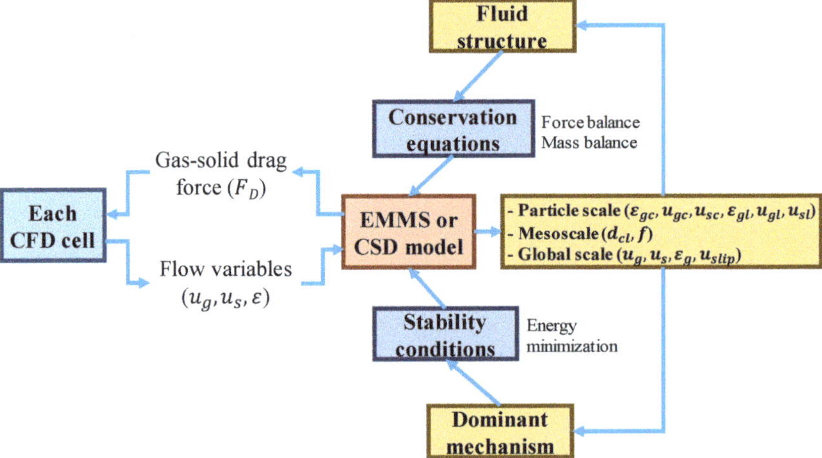

Figure 5. Concurrent coupling strategy of multiscale Eulerian CFD with energy minimization multi-scale (EMMS) or cluster structure-dependent (CSD) in fluidized beds (modified from Li et al. (2018) [87,100]).

3.5.2. Sequential Coupling Strategy

Raynal and Royon-Lebeaud (2007) [101] developed a typical sequential multiscale CFD approach from small to large scales of structured-packing columns for achieving an optimum column design. A 2D VOF-CFD model was used to calculate the liquid holdup and velocity at a small scale of the structured-packing. A gas-phase 3D CFD model with a moving wall boundary condition was applied to an REV of the structured-packing to calculate the relationship between the pressure drop and the gas superficial velocity. Once the liquid holdup, gas-liquid interaction, and pressure drop were determined at the small scales, a gas-solid 3D porous CFD simulation was performed for the whole absorber to examine the influence of internals (e.g., gas and liquid distributors) on the hydrodynamics of the absorber. As a direct CFD simulation that completely considers the complex geometry of structured packing requires significant computational time, the three-step multiscale approach is useful for investigating the hydrodynamics of industrial-scale absorbers.

Qi et al. (2017) [102] proposed a sequential multiscale CFD method to predict the liquid distribution in Mellapak 350Y structured packing. A microscale VOF-CFD model was solved for six REVs at different positions of the packing. Using a macroscale unit network model (UNM), the liquid distribution at the macroscale was evaluated from the distribution coefficient computed by the microscale VOF-CFD model. The macroscale UNM assisted in the design and optimization of structured packing [102].

Ngo et al. (2017) [29] presented the multiscale CFD model of an impregnation die for carbon fiber (CF) prepreg production. A liquid-phase Eulerian CFD was used at the microscale tow domain to obtain resin permeability (K). A liquid-phase steady-state Eulerian CFD with K was applied to the porous media tow domain at the macroscale. Figure 6 shows a sequential coupling strategy between the two Eulerian CFD models. The resin penetration velocity (u_p) at the two tow scales was interconnected, therefore several iterations were necessary to obtain a converged K value.

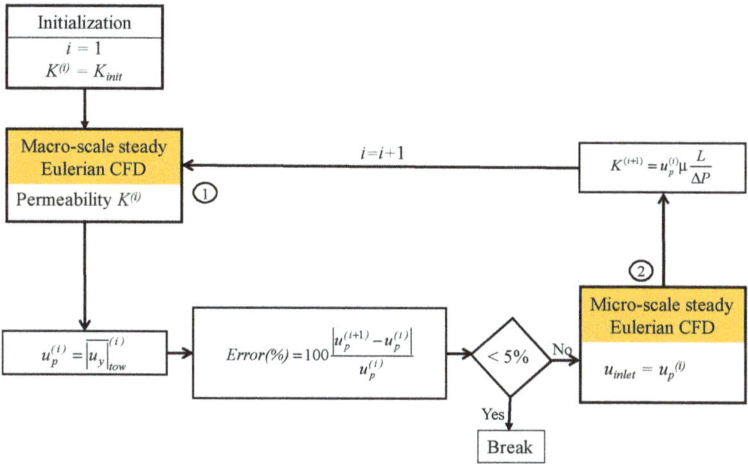

Figure 6. Sequential coupling strategy of multiscale CFD simulation to predict the permeability of resin through porous tows for the production of thermoplastic carbon fibers (modified from Ngo et al. (2017) [29]).

Ngo et al. (2018) [33] proposed another multiscale simulation approach to predict the degree of impregnation in the CF prepreg production process (see Figure 1). The multiscale simulation included three scales: a 2D unsteady-state VOF-CFD for an REV of the microscopic tow domain (see Figure 1d), a liquid-phase Eulerian 3D CFD for the entire impregnation die (see Figure 1b), and a process-scale simulation combining data from the micro- and macroscale CFDs. Figure 7 shows a sequential coupling strategy between the three scales.

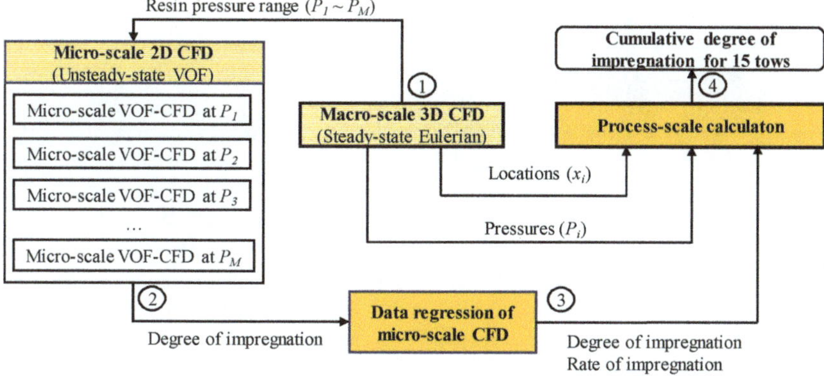

Figure 7. Sequential coupling strategy of multiscale CFD simulation to predict the degree of resin impregnation for the production of thermoplastic carbon fiber (modified from Ngo et al. (2018) [33]).

In gas-solid fluidization systems, several multiscale concepts for the Eulerian CFD model have been proposed: a filtered drag force (FDF) model using constitutive equations from highly resolved subgrid models [103], a mesoscale structure-based drag model, where the flow structure and stability conditions were considered to capture the clustering behavior of solid particles using the EMMS model [46,73,104,105], and a CSD drag model [87,100].

Lu et al. (2019) and Wang (2020) [46,47] reviewed several sequential coupling strategies between the EMMS drag and Eulerian CFD models to speed up the multiscale simulation using a lookup table [105] or a fitting function [104] of the drag force. Figure 8 shows the sequential coupling strategy of the multiscale Eulerian CFD with EMMS in fluidized beds. The flow regime with superficial velocity (u_g) and solid flux (G_s), solid material property with particles density (ρ_p) and diameter (d_p), and a broad range of flow conditions with the Reynolds number (Re) and voidage (ε) are used as the input values of the EMMS model for generating a fitting function of the heterogeneity index (H_D) that is directly multiplied by the original Wen and Yu drag force [104].

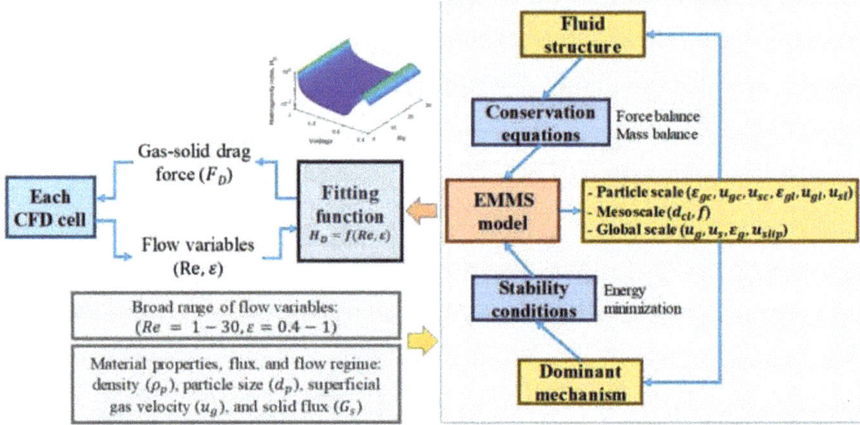

Figure 8. Sequential coupling strategy of a multiscale Eulerian CFD with EMMS in fluidized beds (modified from Lu et al. (2009) [104,105]).

4. Discussion of Multiscale CFD Simulations

In recent CFD studies, multiphase, multiphysics, and multiscale simulations with complex geometries were performed with the increase in computational capacity. Using CFD models validated with experimental data, efforts to replace the experiments are continuing to reduce the time and cost required during process development. However, current CFD simulations are still far from meeting practical demands for process development.

The main challenges limiting the use of CFD simulations are accuracy and computational cost [17]. Despite the successful spread and usage of CFD simulations, there are potential risks of its misuse because of poor model choice, bad assumptions, and wrong interpretation of the results [39]. Reliable theories and numerical methods are needed for accurate simulation. Particular emphasis is placed on the validation of CFD models via experiments and suitable analytic solutions. Extensive validation with experimental data is still required for closure models under a wide range of flow conditions [51].

Table 6 summarizes the recent progress of multiscale CFD simulations in the concurrent and sequential coupling frameworks. Macroscale CFD models have been coupled with LBM, KMC, DEM, EMMS, CSD, and PBE at smaller scales to more accurately calculate physical phenomena at the cost of computational time. In the sequential coupling approach, model parameters required in the macroscale CFD were evaluated at smaller scales or vice versa.

Table 6. Recent progress of multiscale CFD simulation of chemical and biological processes.

	Advantage	Disadvantage	Author	Models	Application Field
Concurrent coupling	High accuracy	High computational cost	Chen, 2013 [98]	LBM-CFD	PEM fuel cell
			Crose, 2017 [13]	KMC-CFD	PECVD wafer
			Pozzetti, 2018 [99]	DEM-VOF	Three-phase flows with particles
			Da Rosa, 2018 [34]	PBE-CFD	Tubular crystallizer
			Qi, 2007 [100]	EMMS-CFD	Fluidized beds
			Li, 2018 [87]	CSD-CFD	Fluidized beds
Sequential coupling	Low accuracy	Low computational cost	Raynal, 2007 [101]	VOF-EM-CFD	Absorber with structured packing
			Ngo, 2017 [29]	Eulerian CFD	Impregnation die for carbon fiber
			Ngo, 2018 [33]	VOF-CFD	Impregnation die for carbon fiber
			Qi, 2017 [102]	VOF-UNM	Absorber with structure packing
			Wang, 2007 [105]	EMMS-CFD	Fluidized beds
			Lu, 2009 [104]	EMMS-CFD	Fluidized beds

Many current works on multiscale modeling focuses on concurrent coupling [4]. Several libraries have been presented to facilitate the concurrent coupling effort for independently developed solvers. Bernaschi et al. (2009) [106] presented a multi-physics/scale code (MUPHY) based on the combination of a microscopic MD with a macroscopic LBM, using MPI as the communication interface for parallelization. Tang et al. (2015) [107] reported a multiscale universal interface (MUI) which is capable of creating an easily customizable framework for solver-dependent data interpretation. Neumann et al. (2016) [108] developed a macro-micro-coupling tool (MaMiCo) for the modulation of molecular-continuum simulations in fluid dynamics. The computational domain was split into a continuum region and a molecular region. At the interface between the two regions, flow quantities such as density, velocity, and temperature were exchanged and matched. MaMiCo supports 2D and 3D simulations, interface definitions for arbitrary MD and CFD codes, MPI-parallel execution, and time-dependent simulations [109].

In addition to multiscale CFD simulations, the mesh quality in 2D or 3D geometries is a prerequisite for achieving accurate CFD simulations. An open-source package, OpenFOAM, based on FVM, has been widely used in CFD because one can customize solvers and easily extend the numerical libraries [110]. NASA defined digital twins (DT) as an integrated multiphysics, multiscale, probabilistic simulation of a system that uses the best available physical models and sensor updates to mirror the life of its flying twin [111]. Simulation tools such as CFD are a key aspect related to DT [112]. This section presents the current research on mesh independency, OpenFOAM, and DT.

4.1. Mesh Independency

CFD models are solved numerically on meshes discretized in a 2D or 3D computational domain. The quality of the mesh plays an important role in the accuracy and stability of the numerical computation [22,113]. The discretization error vanishes only if the grid spacing tends toward zero [114]. A proper number of meshes should be determined as a compromise between computation efficiency and accuracy [68]. Thus, a mesh independence test should be performed in the verification step of the CFD calculation [115].

In a simple mesh independence test, three mesh systems—coarse, medium, and fine—were used [53,54,68,75]. The mesh number of the finer mesh system was double or triple that of the coarser mesh system. To ensure good mesh quality, the minimum orthogonal index was more than 0.01, and the average value of the orthogonal index was significantly higher than 0.01. The maximum skewness index was maintained below 0.95. In addition, the maximum aspect ratio was less than 100, which is much less for flow exhibiting strong gradients [22]. Hydrodynamic values such pressure, velocity, phase fraction, and temperature were compared with respect to the three mesh systems. The medium mesh system was selected, considering the accuracy of the CFD solution and the computation cost [22,54,68,116].

Based on the Richardson extrapolation theory, Roache (1994) [114] proposed a grid convergence index (GCI) to evaluate the numerical uncertainties of the CFD model. The GCI between two grid systems was evaluated. When the ratio of the two GCIs was close to unity, a proper mesh system was selected [115].

For gas-solid fluidized beds, Cloete et al. (2016) [117] investigated the grid independence behavior via the particle relaxation time for several common drag models. The practical rule for the hydrodynamic grid size (Δ_{hydro}) expressed by a function of slip velocity (u_{slip}) was found as [117]

$$\Delta_{hydro} \approx 0.0064 u_{slip} \tag{5}$$

The required grid size for a reactive fluidized-bed (Δ_{react}) was also proposed [117].

$$\Delta_{react} \approx 0.0064 u_{slip} \left(\frac{\frac{\left(\sqrt{27.2^2 + 0.0408Ar} - 27.2 \right)\mu_g}{d_p \rho_s}}{0.012 k_{emul}} \right) \tag{6}$$

where the μ_g, d_p, ρ_s, ρ_g, k_{emul}, and $Ar = \frac{g d_p^3 \rho_g (\rho_s - \rho_g)}{\mu_g^2}$ are the gas viscosity, particle diameter, solid-phase density, gas-phase density, bubble-to-emulsion mass transfer coefficient, and Archimedes number, respectively.

4.2. Open Source Code

In an open-source CFD such as OpenFOAM (open source field operation and manipulation), the user can access the source code and customize it to suit the user's needs [71,118]. Bhusare et al. (2017) [88] investigated the effect of internals on hydrodynamics in a bubble column using a gas-liquid Eulerian CFD model in OpenFOAM, which was compared with experiments and ANSYS Fluent (as a commercial CFD code). Kone et al. (2018) developed an OpenFOAM code to calculate a single-phase 3D CFD model in a PEM fuel cell [71]. Da Rosa and Braatz (2018) [34] presented a multiscale CFD model combined with a micromixing, energy balance and PBE in a tubular crystallizer using OpenFOAM. Farias et al. (2019) presented a liquid-solid Eulerian CFD model coupled with KTGF and PBE to simulate a lovastatin/methanol crystallizer using OpenFOAM [49]. The number of publications using OpenFOAM is increasing.

4.3. Digital Twins with CFD

A reliable CFD model is required to obtain meaningful results from the CFD [26], as mentioned previously. Wang and Xiao (2016) [119] proposed a data-driven CFD modeling of turbulent flows. The drag force (F_D) for a porous medium domain, which was involved in a liquid-phase Eulerian CFD model, was inferred by assimilating experimental data (e.g., velocity and turbulence kinetic energy) and the CFD model prediction in an iterative ensemble Kalman method as an inversion approach [119]. If the experimental data are measured in real-time and provided to the CFD, a virtual machine realized by the CFD can be synchronized with the real machine (experimental apparatus). The main feature of DT is the real-time communication between the real and virtual machines, providing reliable CFD results. However, the real and virtual machines may be different from each other because errors can come from many sources, including a numerical discretization error or a modeling error. The computational cost increases owing to the iterative ensemble method [119].

Lu et al. (2020) [111] reviewed the recent development of DT for smart manufacturing in the context of Industry 4.0. DT, acting as a mirror of the real world, provides a smart tool for simulating, predicting, and optimizing physical manufacturing processes. Studies on standards, communication protocols, time-sensitive data processing, and the reliability of models need to be priorities for the next stage of research in DT [111].

5. Conclusions

This review explored exclusively Eulerian CFD simulations of chemical and biological processes at the multiscale in time and space. To shed light on the progress of CFD technologies in the past decade,

CFD theories and their applications were reviewed in single and multiple phases. The following are the key conclusions from the review:

(1) Chemical and biological processes have multiscale, multiphysics, and multiphase features in nature.

(2) The Eulerian CFD model considering fluid as a continuum includes the volume of fluid (VOF) model, the Eulerian multiphase (EM) model, and the kinetic theory of granular flow (KTGF).

(3) The Eulerian CFD has been used for designing, scaling up, and optimizing at the process-level. Moreover, the Eulerian CFD is suitable for investigating the hydrodynamics (e.g., pressure, velocity, and volume fraction) of a process following geometrical and operational modifications.

(4) RANS-based turbulence equations such as k-ε two-equation types are often used as a general-purpose design tool. The direct numerical simulation (DNS) and large eddy simulation (LES) may be impractical for most industrial flow conditions because of computational time.

(5) The Eulerian CFD has been applied to chemical and biological processes such as gas distributors, combustors, gas storage tanks, bioreactors, fuel cells, random and structured-packing columns, gas-liquid bubble columns, and gas-solid and gas-liquid-solid fluidized beds.

(6) The critical task in multiscale modeling is to exchange information at the interface of neighboring sub-domains. The multiscale CFD strategy interacting between micro- and macroscale domains can be divided into concurrent and sequential couplings. Concurrent coupling is a powerful approach, but it is more computationally prohibitive than the sequential approach.

(7) In the concurrent coupling framework, the Eulerian CFD at a macroscale is solved simultaneously with a lattice Boltzmann method (LBM), kinetic Monte Carlo (KMC), or discrete element model (DEM) at a micro- or mesoscale, updating information at each time step. In the sequential coupling framework, several parameters required for the Eulerian CFD model are calculated from a VOF or EMMS model at a micro- or mesoscale. When the model parameters are interconnected at the two scales, several iterations are necessary to obtain a converged parameter value.

(8) Modeling, meshing, physical properties, and selection of a suitable turbulence model play an important role in CFD simulations. In particular, the proper number of meshes should be determined as a compromise between computation efficiency and accuracy. The mesh independence test is often performed to determine a proper number of meshes in the verification step of the CFD calculation.

(9) Open-source packages such as OpenFOAM are widely used in CFD because one can customize solvers and easily extend the numerical libraries.

(10) CFD models have to be validated by comparing the simulation results with experimental data or suitable analytic solutions to provide meaningful results. If the experimental data are measured in real-time and provided to the CFD, a virtual machine realized by the CFD can be synchronized with the real machine (experimental apparatus). The main feature of digital twins (DT) is the real-time communication between the real and virtual machines, providing reliable CFD results. DT acting as a mirror of the real world provides a smart tool for simulating, predicting, and optimizing physical processes.

(11) The multiscale methods are attractive for industrial applications. However, substantial efforts in physical modeling and numerical implementation are still required before their widespread implementation. A powerful parallelized computational capacity is needed for achieving multiscale simulation and DT.

Author Contributions: Conceptualization, Y.-I.L. and S.I.N.; methodology, S.I.N.; software, S.I.N.; validation, S.I.N. and Y.-I.L.; formal analysis, S.I.N.; investigation, Y.-I.L. and S.I.N.; resources, S.I.N.; data curation, S.I.N.; writing—original draft preparation, S.I.N.; writing—review and editing, Y.-I.L.; visualization, S.I.N.; supervision, Y.-I.L.; project administration, Y.-I.L.; funding acquisition, Y.-I.L. All authors have read and agreed to the published version of the manuscript.

Funding: This research was funded by the Korea government (MSIT), grant number: NRF-2019R1H1A2079924, and the Ministry of Science, ICT and Future Planning (MSIP), grant number: CRC-14-1-KRICT.

Acknowledgments: This research was supported by the Basic Science Research Program through the National Research Foundation of Korea (NRF) and the National Research Council of Science & Technology (NST) of Korea.

Conflicts of Interest: The authors declare no conflict of interest. The funders had no role in the design of the study; in the collection, analyses, or interpretation of data; in the writing of the manuscript, or in the decision to publish the results.

References

1. Delgado-Buscalioni, R.; Coveney, P.V.; Riley, G.D.; Ford, R.W. Hybrid molecular-continuum fluid models: Implementation within a general coupling framework. *Philos. Trans. R. Soc. A* **2005**, *363*, 1975–1985. [CrossRef]
2. Lim, Y.-I. State-of-arts in multiscale simulation for process development. *Korean Chem. Eng. Res.* **2013**, *51*, 10–24. [CrossRef]
3. Drikakis, D.; Frank, M.; Tabor, G. Multiscale computational fluid dynamics. *Energies* **2019**, *12*, 3272. [CrossRef]
4. Weinan, E. *Principles of Multiscale Modeling*, 1st ed.; Cambridge University Press: New York, NY, USA, 2011.
5. Mahian, O.; Kolsi, L.; Amani, M.; Estellé, P.; Ahmadi, G.; Kleinstreuer, C.; Marshall, J.S.; Taylor, R.A.; Abu-Nada, E.; Rashidi, S.; et al. Recent advances in modeling and simulation of nanofluid flows—Part II: Applications. *Phys. Rep.* **2019**, *791*, 1–59. [CrossRef]
6. Braatz, R.D. Multiscale simulation in science and engineering. In Proceedings of the AIChE Annual Meeting, Nashville, TN, USA, 8–13 November 2009.
7. Fermeglia, M.; Pricl, S. Multiscale molecular modeling in nanostructured material design and process system engineering. *Comput. Chem. Eng.* **2009**, *33*, 1701–1710. [CrossRef]
8. Olsen, J.M.H.; Bolnykh, V.; Meloni, S.; Ippoliti, E.; Bircher, M.P.; Carloni, P.; Rothlisberger, U. MiMiC: A novel framework for multiscale modeling in computational chemistry. *J. Chem. Theory Comput.* **2019**, *15*, 3810–3823. [CrossRef] [PubMed]
9. Ideker, T.; Galitski, T.; Hood, L. A new approach to decoding life: Systems biology. *Annu. Rev. Genom. Hum. Genet.* **2001**, *2*, 343–372. [CrossRef] [PubMed]
10. Renardy, M.; Hult, C.; Evans, S.; Linderman, J.J.; Kirschner, D.E. Global sensitivity analysis of biological multiscale models. *Curr. Opin. Biomed. Eng.* **2019**, *11*, 109–116. [CrossRef]
11. Durdagi, S.; Dogan, B.; Erol, I.; Kayık, G.; Aksoydan, B. Current status of multiscale simulations on GPCRs. *Curr. Opin. Struct. Biol.* **2019**, *55*, 93–103. [CrossRef]
12. Koullapis, P.; Ollson, B.; Kassinos, S.C.; Sznitman, J. Multiscale in silico lung modeling strategies for aerosol inhalation therapy and drug delivery. *Curr. Opin. Biomed. Eng.* **2019**, *11*, 130–136. [CrossRef]
13. Crose, M.; Tran, A.; Christofides, D.P. Multiscale computational fluid dynamics: Methodology and application to PECVD of thin film solar cells. *Coatings* **2017**, *7*, 22. [CrossRef]
14. Raimondeau, S.; Vlachos, D.G. Recent developments on multiscale, hierarchical modeling of chemical reactors. *Chem. Eng. J.* **2002**, *90*, 3–23. [CrossRef]
15. Tong, Z.-X.; He, Y.-L.; Tao, W.-Q. A review of current progress in multiscale simulations for fluid flow and heat transfer problems: The frameworks, coupling techniques and future perspectives. *Int. J. Heat Mass Transf.* **2019**, *137*, 1263–1289. [CrossRef]
16. Vlachos, D.G. Multiscale modeling for emergent behavior, complexity, and combinatorial explosion. *AIChE J.* **2012**, *58*, 1314–1325. [CrossRef]
17. Ge, W.; Chang, Q.; Li, C.; Wang, J. Multiscale structures in particle–fluid systems: Characterization, modeling, and simulation. *Chem. Eng. Sci.* **2019**, *198*, 198–223. [CrossRef]
18. Mahian, O.; Kolsi, L.; Amani, M.; Estellé, P.; Ahmadi, G.; Kleinstreuer, C.; Marshall, J.S.; Siavashi, M.; Taylor, R.A.; Niazmand, H.; et al. Recent advances in modeling and simulation of nanofluid flows-Part I: Fundamentals and theory. *Phys. Rep.* **2019**, *790*, 1–48. [CrossRef]
19. Chen, S.; Doolen, G.D. Lattice Boltzmann method for fluid flows. *Annu. Rev. Fluid Mech.* **1998**, *30*, 329–364. [CrossRef]
20. Nourgaliev, R.R.; Dinh, T.N.; Theofanous, T.G.; Joseph, D. The lattice Boltzmann equation method: Theoretical interpretation, numerics and implications. *Int. J. Multiph. Flow* **2003**, *29*, 117–169. [CrossRef]
21. Nguyen, T.D.B.; Seo, M.W.; Lim, Y.-I.; Song, B.-H.; Kim, S.-D. CFD simulation with experiments in a dual circulating fluidized bed gasifier. *Comput. Chem. Eng.* **2012**, *36*, 48–56. [CrossRef]

22. Nguyen, D.D.; Ngo, S.I.; Lim, Y.-I.; Kim, W.; Lee, U.-D.; Seo, D.; Yoon, W.-L. Optimal design of a sleeve-type steam methane reforming reactor for hydrogen production from natural gas. *Int. J. Hydrog. Energy* **2019**, *44*, 1973–1987. [CrossRef]

23. Pan, H.; Chen, X.-Z.; Liang, X.-F.; Zhu, L.-T.; Luo, Z.-H. CFD simulations of gas–liquid–solid flow in fluidized bed reactors—A review. *Powder Technol.* **2016**, *299*, 235–258. [CrossRef]

24. Pham, H.H.; Lim, Y.-I.; Han, S.; Lim, B.; Ko, H.-S. Hydrodynamics and design of gas distributor in large-scale amine absorbers using computational fluid dynamics. *Korean J. Chem. Eng.* **2018**, *35*, 1073–1082. [CrossRef]

25. Yang, N.; Xiao, Q. A mesoscale approach for population balance modeling of bubble size distribution in bubble column reactors. *Chem. Eng. Sci.* **2017**, *170*, 241–250. [CrossRef]

26. Bourgeois, T.; Ammouri, F.; Baraldi, D.; Moretto, P. The temperature evolution in compressed gas filling processes: A review. *Int. J. Hydrog. Energy* **2018**, *43*, 2268–2292. [CrossRef]

27. Raveh, D.E. Computational-fluid-dynamics-based aeroelastic analysis and structural design optimization—A researcher's perspective. *Comput. Methods Appl. Mech. Eng.* **2005**, *194*, 3453–3471. [CrossRef]

28. Pinto, G.; Silva, F.; Porteiro, J.; Míguez, J.; Baptista, A. Numerical simulation applied to PVD reactors: An overview. *Coatings* **2018**, *8*, 410. [CrossRef]

29. Ngo, S.I.; Lim, Y.-I.; Hahn, M.-H.; Jung, J.; Bang, Y.-H. Multi-scale computational fluid dynamics of impregnation die for thermoplastic carbon fiber prepreg production. *Comput. Chem. Eng.* **2017**, *103*, 58–68. [CrossRef]

30. Lakehal, D. Status and future developments of Large-Eddy Simulation of turbulent multi-fluid flows (LEIS and LESS). *Int. J. Multiph. Flow* **2018**, *104*, 322–337. [CrossRef]

31. Xiao, H.; Cinnella, P. Quantification of model uncertainty in RANS simulations: A review. *Prog. Aerosp. Sci.* **2019**, *108*, 1–31. [CrossRef]

32. Klein, M.; Ketterl, S.; Hasslberger, J. Large eddy simulation of multiphase flows using the volume of fluid method: Part 1—Governing equations and a priori analysis. *Exp. Comput. Multiph. Flow* **2019**, *1*, 130–144. [CrossRef]

33. Ngo, S.I.; Lim, Y.-I.; Hahn, M.-H.; Jung, J. Prediction of degree of impregnation in thermoplastic unidirectional carbon fiber prepreg by multi-scale computational fluid dynamics. *Chem. Eng. Sci.* **2018**, *185*, 64–75. [CrossRef]

34. da Rosa, C.A.; Braatz, R.D. Multiscale modeling and simulation of macromixing, micromixing, and crystal size distribution in radial mixers/crystallizers. *Ind. Eng. Chem. Res.* **2018**, *57*, 5433–5441. [CrossRef]

35. Haghighat, A.; Luxbacher, K.; Lattimer, B.Y. Development of a methodology for interface boundary selection in the multiscale road tunnel fire simulations. *Fire Technol.* **2018**, *54*, 1029–1066. [CrossRef]

36. Höhne, T.; Krepper, E.; Lucas, D.; Montoya, G. A multiscale approach simulating boiling in a heated pipe including flow pattern transition. *Nucl. Technol.* **2019**, *205*, 48–56. [CrossRef]

37. Uribe, S.; Cordero, M.E.; Reyes, E.P.; Regalado-Méndez, A.; Zárate, L.G. Multiscale CFD modelling and analysis of TBR behavior for an HDS process: Deviations from ideal behaviors. *Fuel* **2019**, *239*, 1162–1172. [CrossRef]

38. Ferreira, R.B.; Falcão, D.S.; Oliveira, V.B.; Pinto, A.M.F.R. Numerical simulations of two-phase flow in proton exchange membrane fuel cells using the volume of fluid method—A review. *J. Power Sources* **2015**, *277*, 329–342. [CrossRef]

39. Karpinska, A.M.; Bridgeman, J. CFD-aided modelling of activated sludge systems—A critical review. *Water Res.* **2016**, *88*, 861–879. [CrossRef]

40. Sharma, S.K.; Kalamkar, V.R. Computational fluid dynamics approach in thermo-hydraulic analysis of flow in ducts with rib roughened walls—A review. *Renew. Sustain. Energy Rev.* **2016**, *55*, 756–788. [CrossRef]

41. Yin, C.; Yan, J. Oxy-fuel combustion of pulverized fuels: Combustion fundamentals and modeling. *Appl. Energy* **2016**, *162*, 742–762. [CrossRef]

42. Uebel, K.; Rößger, P.; Prüfert, U.; Richter, A.; Meyer, B. CFD-based multi-objective optimization of a quench reactor design. *Fuel Process. Technol.* **2016**, *149*, 290–304. [CrossRef]

43. Fotovat, F.; Bi, X.T.; Grace, J.R. Electrostatics in gas-solid fluidized beds: A review. *Chem. Eng. Sci.* **2017**, *173*, 303–334. [CrossRef]

44. Pires, J.C.M.; Alvim-Ferraz, M.C.M.; Martins, F.G. Photobioreactor design for microalgae production through computational fluid dynamics: A review. *Renew. Sustain. Energy Rev.* **2017**, *79*, 248–254. [CrossRef]

45. Malekjani, N.; Jafari, S.M. Simulation of food drying processes by Computational Fluid Dynamics (CFD); recent advances and approaches. *Trends Food Sci. Technol.* **2018**, *78*, 206–223. [CrossRef]

46. Lu, B.; Niu, Y.; Chen, F.; Ahmad, N.; Wang, W.; Li, J. Energy-minimization multiscale based mesoscale modeling and applications in gas-fluidized catalytic reactors. *Rev. Chem. Eng.* **2019**. [CrossRef]

47. Wang, J. Continuum theory for dense gas-solid flow: A state-of-the-art review. *Chem. Eng. Sci.* **2020**, *215*, 115428. [CrossRef]

48. Pham, H.H.; Lim, Y.-I.; Ngo, S.I.; Bang, Y.-H. Computational fluid dynamics and tar formation in a low-temperature carbonization furnace for the production of carbon fibers. *J. Ind. Eng. Chem.* **2019**, *73*, 286–296. [CrossRef]

49. Farias, L.F.I.; de Souza, J.A.; Braatz, R.D.; da Rosa, C.A. Coupling of the population balance equation into a two-phase model for the simulation of combined cooling and antisolvent crystallization using OpenFOAM. *Comput. Chem. Eng.* **2019**, *123*, 246–256. [CrossRef]

50. Ngo, S.I.; Lim, Y.-I.; Kim, S.-C. Wave characteristics of coagulation bath in dry-jet wet-spinning process for polyacrylonitrile fiber production using computational fluid dynamics. *Processes* **2019**, *7*, 314. [CrossRef]

51. Wu, B.; Firouzi, M.; Mitchell, T.; Rufford, T.E.; Leonardi, C.; Towler, B. A critical review of flow maps for gas-liquid flows in vertical pipes and annuli. *Chem. Eng. J.* **2017**, *326*, 350–377. [CrossRef]

52. Bhole, M.R.; Joshi, J.B.; Ramkrishna, D. CFD simulation of bubble columns incorporating population balance modeling. *Chem. Eng. Sci.* **2008**, *63*, 2267–2282. [CrossRef]

53. Tran, B.V.; Nguyen, D.D.; Ngo, S.I.; Lim, Y.-I.; Kim, B.; Lee, D.H.; Go, K.-S.; Nho, N.-S. Hydrodynamics and simulation of air–water homogeneous bubble column under elevated pressure. *AIChE J.* **2019**, *65*, e16685. [CrossRef]

54. Pham, D.A.; Lim, Y.-I.; Jee, H.; Ahn, E.; Jung, Y. Porous media Eulerian computational fluid dynamics (CFD) model of amine absorber with structured-packing for CO_2 removal. *Chem. Eng. Sci.* **2015**, *132*, 259–270. [CrossRef]

55. Ngo, S.I.; Lim, Y.-I.; Song, B.-H.; Lee, U.-D.; Lee, J.-W.; Song, J.-H. Effects of fluidization velocity on solid stack volume in a bubbling fluidized-bed with nozzle-type distributor. *Powder Technol.* **2015**, *275*, 188–198. [CrossRef]

56. Ngo, S.I.; Lim, Y.-I.; Song, B.-H.; Lee, U.-D.; Yang, C.-W.; Choi, Y.-T.; Song, J.-H. Hydrodynamics of cold-rig biomass gasifier using semi-dual fluidized-bed. *Powder Technol.* **2013**, *234*, 97–106. [CrossRef]

57. Sun, L.; Luo, K.; Fan, J. Numerical investigation on methanation kinetic and flow behavior in full-loop fluidized bed reactor. *Fuel* **2018**, *231*, 85–93. [CrossRef]

58. Zhou, Q.; Wang, J. CFD study of mixing and segregation in CFB risers: Extension of EMMS drag model to binary gas–solid flow. *Chem. Eng. Sci.* **2015**, *122*, 637–651. [CrossRef]

59. Gidaspow, D. *Multiphase Flow and Fluidization: Continuum and Kinetic Theory Description*; Academic Press: Cambridge, MA, USA, 1994.

60. Lun, C.K.K.; Savage, S.B.; Jeffrey, D.J.; Chepurniy, N. Kinetic theories for granular flow: Inelastic particles in Couette flow and slightly inelastic particles in a general flowfield. *J. Fluid Mech.* **1984**, *140*, 223–256. [CrossRef]

61. Syamlal, M.; Rogers, W.; O'Brien, T.J. *MFIX Documentation: Volume1, Theory Guide*; National Technical Information Service: Springfield, VA, USA, 1993.

62. Schaeffer, D.G. Instability in the evolution equations describing incompressible granular flow. *J. Differ. Equ.* **1987**, *66*, 19–50. [CrossRef]

63. Ogawa, S.; Umemura, A.; Oshima, N. On the equations of fully fluidized granular materials. *Z. Angew. Math. Phys. ZAMP* **1980**, *31*, 483–493. [CrossRef]

64. Ge, W.; Wang, W.; Yang, N.; Li, J.; Kwauk, M.; Chen, F.; Chen, J.; Fang, X.; Guo, L.; He, X.; et al. Meso-scale oriented simulation towards virtual process engineering (VPE)-The EMMS Paradigm. *Chem. Eng. Sci.* **2011**, *66*, 4426–4458. [CrossRef]

65. Hamidipour, M.; Chen, J.; Larachi, F. CFD study on hydrodynamics in three-phase fluidized beds—Application of turbulence models and experimental validation. *Chem. Eng. Sci.* **2012**, *78*, 167–180. [CrossRef]

66. Hosseini, S.H.; Shojaee, S.; Ahmadi, G.; Zivdar, M. Computational fluid dynamics studies of dry and wet pressure drops in structured packings. *J. Ind. Eng. Chem.* **2012**, *18*, 1465–1473. [CrossRef]

67. Nguyen, T.D.B.; Lim, Y.-I.; Kim, S.-J.; Eom, W.-H.; Yoo, K.-S. Experiment and Computational Fluid Dynamics (cfd) simulation of urea-based Selective Noncatalytic Reduction (SNCR) in a pilot-scale flow reactor. *Energy Fuels* **2008**, *22*, 3864–3876. [CrossRef]
68. Ngo, S.I.; Lim, Y.-I.; Kim, W.; Seo, D.J.; Yoon, W.L. Computational fluid dynamics and experimental validation of a compact steam methane reformer for hydrogen production from natural gas. *Appl. Energy* **2019**, *236*, 340–353. [CrossRef]
69. Rzehak, R.; Krauß, M.; Kováts, P.; Zähringer, K. Fluid dynamics in a bubble column: New experiments and simulations. *Int. J. Multiph. Flow* **2017**, *89*, 299–312. [CrossRef]
70. Ramos, A.; Monteiro, E.; Rouboa, A. Numerical approaches and comprehensive models for gasification process: A review. *Renew. Sustain. Energy Rev.* **2019**, *110*, 188–206. [CrossRef]
71. Kone, J.-P.; Zhang, X.; Yan, Y.; Hu, G.; Ahmadi, G. CFD modeling and simulation of PEM fuel cell using OpenFOAM. *Energy Procedia* **2018**, *145*, 64–69. [CrossRef]
72. Ahmadi, S.; Sefidvash, F. Study of pressure drop in fixed bed reactor using a Computational Fluid Dynamics (CFD) code. *Chem. Eng.* **2018**, *2*, 14.
73. Wang, J.-Q.; Ouyang, Y.; Li, W.-L.; Esmaeili, A.; Xiang, Y.; Chen, J.-F. CFD analysis of gas flow characteristics in a rotating packed bed with randomly arranged spherical packing. *Chem. Eng. J.* **2020**, *385*, 123812. [CrossRef]
74. Kang, J.-L.; Ciou, Y.-C.; Lin, D.-Y.; Wong, D.S.-H.; Jang, S.-S. Investigation of hydrodynamic behavior in random packing using CFD simulation. *Chem. Eng. Res. Des.* **2019**, *147*, 43–54. [CrossRef]
75. Kim, J.; Pham, D.A.; Lim, Y.-I. Effect of gravity center position on amine absorber with structured packing under offshore operation: Computational fluid dynamics approach. *Chem. Eng. Res. Des.* **2017**, *121*, 99–112. [CrossRef]
76. Soman, A.; Shastri, Y. Optimization of novel photobioreactor design using computational fluid dynamics. *Appl. Energy* **2015**, *140*, 246–255. [CrossRef]
77. Le, T.T.; Ngo, S.I.; Lim, Y.-I.; Park, C.-K.; Lee, B.-D.; Kim, B.-G.; Lim, D.-H. Effect of simultaneous three-angular motion on the performance of an air–water–oil separator under offshore operation. *Ocean Eng.* **2019**, *171*, 469–484. [CrossRef]
78. Fletcher, D.F.; McClure, D.D.; Kavanagh, J.M.; Barton, G.W. CFD simulation of industrial bubble columns: Numerical challenges and model validation successes. *Appl. Math. Model.* **2017**, *44*, 25–42. [CrossRef]
79. Sharma, S.L.; Ishii, M.; Hibiki, T.; Schlegel, J.P.; Liu, Y.; Buchanan, J.R. Beyond bubbly two-phase flow investigation using a CFD three-field two-fluid model. *Int. J. Multiph. Flow* **2019**, *113*, 1–15. [CrossRef]
80. Yan, P.; Jin, H.; He, G.; Guo, X.; Ma, L.; Yang, S.; Zhang, R. CFD simulation of hydrodynamics in a high-pressure bubble column using three optimized drag models of bubble swarm. *Chem. Eng. Sci.* **2019**, *199*, 137–155. [CrossRef]
81. Thakare, H.R.; Monde, A.; Parekh, A.D. Experimental, computational and optimization studies of temperature separation and flow physics of vortex tube: A review. *Renew. Sustain. Energy Rev.* **2015**, *52*, 1043–1071. [CrossRef]
82. Yang, X.; Yu, H.; Wang, R.; Fane, A.G. Optimization of microstructured hollow fiber design for membrane distillation applications using CFD modeling. *J. Membr. Sci.* **2012**, *421–422*, 258–270. [CrossRef]
83. Minea, A.A.; Murshed, S.M.S. A review on development of ionic liquid based nanofluids and their heat transfer behavior. *Renew. Sustain. Energy Rev.* **2018**, *91*, 584–599. [CrossRef]
84. Pham, H.H. Performance Evaluation of Low Temperature Furnace and Gas Distributors Using Computational Fluid Dynamics (CFD). Master Thesis, Hankyong National University, Anseong, Korea, 2018.
85. Kim, J.; Pham, D.A.; Lim, Y.-I. Gas–liquid multiphase computational fluid dynamics (CFD) of amine absorption column with structured-packing for CO_2 capture. *Comput. Chem. Eng.* **2016**, *88*, 39–49. [CrossRef]
86. Zhang, G.; Jiao, K. Multi-phase models for water and thermal management of proton exchange membrane fuel cell: A review. *J. Power Sources* **2018**, *391*, 120–133. [CrossRef]
87. Li, W.-L.; Ouyang, Y.; Gao, X.-Y.; Wang, C.-Y.; Shao, L.; Xiang, Y. CFD analysis of gas–liquid flow characteristics in a microporous tube-in-tube microchannel reactor. *Comput. Fluids* **2018**, *170*, 13–23. [CrossRef]
88. Bhusare, V.H.; Dhiman, M.K.; Kalaga, D.V.; Roy, S.; Joshi, J.B. CFD simulations of a bubble column with and without internals by using OpenFOAM. *Chem. Eng. J.* **2017**, *317*, 157–174. [CrossRef]

89. Behkish, A.; Lemoine, R.; Sehabiague, L.; Oukaci, R.; Morsi, B.I. Gas holdup and bubble size behavior in a large-scale slurry bubble column reactor operating with an organic liquid under elevated pressures and temperatures. *Chem. Eng. J.* **2007**, *128*, 69–84. [CrossRef]

90. Zhang, H.; Yang, G.; Sayyar, A.; Wang, T. An improved bubble breakup model in turbulent flow. *Chem. Eng. J.* **2019**. [CrossRef]

91. Abdelmotalib, H.M.; Youssef, M.A.M.; Hassan, A.A.; Youn, S.B.; Im, I.-T. Heat transfer process in gas–solid fluidized bed combustors: A review. *Int. J. Heat Mass Transf.* **2015**, *89*, 567–575. [CrossRef]

92. Wu, Y.; Shi, X.; Liu, Y.; Wang, C.; Gao, J.; Lan, X. 3D CPFD simulations of gas-solids flow in a CFB downer with cluster-based drag model. *Powder Technol.* **2019**. [CrossRef]

93. Banerjee, S.; Agarwal, R.K. Computational fluid dynamics modeling and simulations of fluidized beds for chemical looping combustion. In *Handbook of Chemical Looping Technology*; Breault, R.W., Ed.; Wiley-VCH: Weinheim, Germany, 2020; pp. 303–332. [CrossRef]

94. Andrews, M.J.; O'Rourke, P.J. The multiphase particle-in-cell (MP-PIC) method for dense particulate flows. *Int. J. Multiph. Flow* **1996**, *22*, 379–402. [CrossRef]

95. Snider, D.M.; Clark, S.M.; O'Rourke, P.J. Eulerian–Lagrangian method for three-dimensional thermal reacting flow with application to coal gasifiers. *Chem. Eng. Sci.* **2011**, *66*, 1285–1295. [CrossRef]

96. Kim, S.H.; Lee, J.H.; Braatz, R.D. Multi-phase particle-in-cell coupled with population balance equation (MP-PIC-PBE) method for multiscale computational fluid dynamics simulation. *Comput. Chem. Eng.* **2020**, *134*, 106686. [CrossRef]

97. Li, Y.; Yang, G.Q.; Zhang, J.P.; Fan, L.S. Numerical studies of bubble formation dynamics in gas–liquid–solid fluidization at high pressures. *Powder Technol.* **2001**, *116*, 246–260. [CrossRef]

98. Chen, L.; Feng, Y.-L.; Song, C.-X.; Chen, L.; He, Y.-L.; Tao, W.-Q. Multi-scale modeling of proton exchange membrane fuel cell by coupling finite volume method and lattice Boltzmann method. *Int. J. Heat Mass Transf.* **2013**, *63*, 268–283. [CrossRef]

99. Pozzetti, G.; Peters, B. A multiscale DEM-VOF method for the simulation of three-phase flows. *Int. J. Multiph. Flow* **2018**, *99*, 186–204. [CrossRef]

100. Qi, H.; Li, F.; Xi, B.; You, C. Modeling of drag with the Eulerian approach and EMMS theory for heterogeneous dense gas-solid two-phase flow. *Chem. Eng. Sci.* **2007**, *62*, 1670–1681. [CrossRef]

101. Raynal, L.; Royon-Lebeaud, A. A multi-scale approach for CFD calculations of gas–liquid flow within large size column equipped with structured packing. *Chem. Eng. Sci.* **2007**, *62*, 7196–7204. [CrossRef]

102. Qi, W.; Guo, K.; Liu, C.; Liu, H.; Liu, B. Liquid distribution and local hydrodynamics of winpak: A multiscale method. *Ind. Eng. Chem. Res.* **2017**, *56*, 15184–15194. [CrossRef]

103. Gao, X.; Li, T.; Sarkar, A.; Lu, L.; Rogers, W.A. Development and validation of an enhanced filtered drag model for simulating gas-solid fluidization of Geldart A particles in all flow regimes. *Chem. Eng. Sci.* **2018**, *184*, 33–51. [CrossRef]

104. Lu, B.; Wang, W.; Li, J. Searching for a mesh-independent sub-grid model for CFD simulation of gas–solid riser flows. *Chem. Eng. Sci.* **2009**, *64*, 3437–3447. [CrossRef]

105. Wang, W.; Li, J. Simulation of gas–solid two-phase flow by a multi-scale CFD approach—Of the EMMS model to the sub-grid level. *Chem. Eng. Sci.* **2007**, *62*, 208–231. [CrossRef]

106. Bernaschi, M.; Melchionna, S.; Succi, S.; Fyta, M.; Kaxiras, E.; Sircar, J.K. MUPHY: A parallel MUlti PHYsics/scale code for high performance bio-fluidic simulations. *Comput. Phys. Commun.* **2009**, *180*, 1495–1502. [CrossRef]

107. Tang, Y.-H.; Kudo, S.; Bian, X.; Li, Z.; Karniadakis, G.E. Multiscale universal interface: A concurrent framework for coupling heterogeneous solvers. *J. Comput. Phys.* **2015**, *297*, 13–31. [CrossRef]

108. Neumann, P.; Flohr, H.; Arora, R.; Jarmatz, P.; Tchipev, N.; Bungartz, H.-J. MaMiCo: Software design for parallel molecular-continuum flow simulations. *Comput. Phys. Commun.* **2016**, *200*, 324–335. [CrossRef]

109. Neumann, P.; Bian, X. MaMiCo: Transient multi-instance molecular-continuum flow simulation on supercomputers. *Comput. Phys. Commun.* **2017**, *220*, 390–402. [CrossRef]

110. Li, L.; Li, B.; Liu, Z. Modeling of spout-fluidized beds and investigation of drag closures using OpenFOAM. *Powder Technol.* **2017**, *305*, 364–376. [CrossRef]

111. Lu, Y.; Liu, C.; Wang, K.I.K.; Huang, H.; Xu, X. Digital twin-driven smart manufacturing: Connotation, reference model, applications and research issues. *Robot. Comput. Integr. Manuf.* **2020**, *61*, 101837. [CrossRef]

112. Negri, E.; Fumagalli, L.; Macchi, M. A review of the roles of digital twin in CPS-based production systems. *Procedia Manuf.* **2017**, *11*, 939–948. [CrossRef]

113. Katz, A.; Sankaran, V. Mesh quality effects on the accuracy of CFD solutions on unstructured meshes. *J. Comput. Phys.* **2011**, *230*, 7670–7686. [CrossRef]

114. Roache, P.J. Perspective: A method for uniform reporting of grid refinement studies. *J. Fluids Eng.* **1994**, *116*, 405–413. [CrossRef]

115. Roache, P.J. Verification of codes and calculations. *AIAA J.* **1998**, *36*, 696–702. [CrossRef]

116. Owens, S.A.; Perkins, M.R.; Eldridge, R.B.; Schulz, K.W.; Ketcham, R.A. Computational fluid dynamics simulation of structured packing. *Ind. Eng. Chem. Res.* **2013**, *52*, 2032–2045. [CrossRef]

117. Cloete, S.; Johansen, S.T.; Amini, S. Grid independence behaviour of fluidized bed reactor simulations using the two fluid model: Detailed parametric study. *Powder Technol.* **2016**, *289*, 65–70. [CrossRef]

118. Liu, Q.; Gómez, F.; Pérez, J.M.; Theofilis, V. Instability and sensitivity analysis of flows using OpenFOAM®. *Chin. J. Aeronaut.* **2016**, *29*, 316–325. [CrossRef]

119. Wang, J.-X.; Xiao, H. Data-driven CFD modeling of turbulent flows through complex structures. *Int. J. Heat Fluid Flow* **2016**, *62*, 138–149. [CrossRef]

Publisher's Note: MDPI stays neutral with regard to jurisdictional claims in published maps and institutional affiliations.

 *chemengineering*

Article

Influence of Pressure, Velocity and Fluid Material on Heat Transport in Structured Open-Cell Foam Reactors Investigated Using CFD Simulations

Christoph Sinn [1], Jonas Wentrup [1], Jorg Thöming [1,2] and Georg R. Pesch [1,2,*]

[1] Faculty of Production Engineering, Chemical Process Engineering Group, University of Bremen, Leobener Strasse 6, 28359 Bremen, Germany; sinn@uni-bremen.de (C.S.); jwentrup@uni-bremen.de (J.W.); thoeming@uni-bremen.de (J.T.)

[2] MAPEX Center for Materials and Processes, University of Bremen, Postbox 330 440, 28334 Bremen, Germany

* Correspondence: gpesch@uni-bremen.de

Received: 4 September 2020; Accepted: 2 November 2020; Published: 14 November 2020

Abstract: Structured open-cell foam reactors are promising for managing highly exothermic reactions such as CO_2 methanation due to their excellent heat transport properties. Especially at low flow rates and under dynamic operation, foam-based reactors can be advantageous over classic fixed-bed reactors. To efficiently design the catalyst carriers, a thorough understanding of heat transport mechanisms is needed. So far, studies on heat transport in foams have mostly focused on the solid phase and used air at atmospheric pressure as fluid phase. With the aid of pore-scale 3d CFD simulations, we analyze the effect of the fluid properties on heat transport under conditions close to the CO_2 methanation reaction for two different foam structures. The exothermicity is mimicked via volumetric uniformly distributed heat sources. We found for foams that are designed to be used as catalyst carriers that the working pressure range and the superficial velocity influence the dominant heat removal mechanism significantly. In contrast, the influence of fluid type and gravity on heat removal is small in the range relevant for heterogeneous catalysis. The findings might help to facilitate the design-process of open-cell foam reactors and to better understand heat transport mechanisms in foams.

Keywords: computational fluid dynamics (CFD); conjugate heat transfer; open-cell foams; structured reactors; volumetric heat sources; fluid properties; STAR-CCM+; dynamic operation

---

1. Introduction

Open-cell foams are promising monolithic catalyst support structures for exo- and endothermal heterogeneous gas-phase reactions such as the CO_2 methanation, due favorable characteristics such as good heat transport, low pressure drop and high porosities [1–3]. The CO_2 methanation currently attracts much attention in the chemical engineering community as it can be a cornerstone the future transition to a sustainable energy supply within the Power-to-X (PtX) concept [4]. Here, excess energy from wind turbines or solar panels is converted via electrolysis into hydrogen, and usually reacted catalytically with carbon dioxide to methane. As the power grids might not be resilient enough to withstand the high energy load during wind or sun peak hours, dynamically operated small-scale plants for storage of excess energy are needed in the future [5]. Small-scale plants are operated at low flow rates making heat removal traditionally difficult. The dynamic operation can cause tremendous jumps of pressure, velocity and even change the fluid composition resulting in undesired or uncontrollable temperature increases. Thus, the heat transport (and heat removal) inside these reactors needs to be addressed to protect the catalyst's activity and ensure safe as well as robust operation [6]. Generally, for the methanation reaction, low temperatures are thermodynamically favorable [7]. Therefore, an effective heat removal ensures efficient conversion and protects the catalyst particles from sintering.

Recently it was shown that, at low flowrates, foam-structured reactors can be advantageous over conventional pellets for CO_2 methanation [8]. The reason is that heat is dominantly removed via radial conduction in foam reactors, whereas pellets in fixed-bed reactors remove heat axially via convection [9]. However, only when heat conduction dominates over convection, foams can be advantageous over pellets as pellets generally have a higher axial convective heat transport [8].

To study heat flows and heat transport mechanisms in structured reactors, computational fluid dynamics (CFD) simulations have been well established [10]. With the aid of CFD simulations, the three heat transport mechanisms, conduction [11,12], convection [13,14] and radiation [15], have been addressed in open-cell foams. Among others, the influences of superficial velocity [16,17], wall coupling [18], solid thermal conductivity [19], geometrical influences (i.e., strut shape [20,21] and porosity [22,23]) on heat transport were studied simulatively. Most of the studies targeted solely heat transport in the solid or conjugate heat transport between a fluid and a foam with no superimposed catalytic reaction. Simulations of detailed catalytic reactions on foams are costly and only a few studies have been published so far that mainly dealt with the CO oxidation reaction [24–28].

In order to reduce complexity of the reaction system and to speed up simulations, we recently proposed to mimic exothermal reactions via uniformly distributed heat sources in the solid foam [19]. Heat sources (or heat sinks) mimic the exothermicity (or endothermicity) and can be used to study heat flows and temperature increases in open-cell foams decoupled from reaction mechanisms. This approach has been utilized to study the influence of heat sources intensities (from 5–150 W), laminar superficial velocities (0–0.5 m s^{-1}), solid thermal conductivities (1–200 W m^{-1} K^{-1}), and thermal radiation (temperature levels 500–1200 K, solid emissivities 0–1) [15,19]. It was found that the solid thermal conductivity and the superficial velocity are key parameters for designing efficient heat-removing foams for the usage as catalyst carriers [19]. Moreover, for reactor tube diameters below 25 mm and superficial velocities lower than 0.5 m s^{-1}, the investigated structure (irregular foam) showed conduction being the dominant heat removal mechanism (this parameter range is in relevant order of magnitude for decentralized small scale methanation plants [29,30]). Furthermore, thermal radiation can contribute significantly to the heat transport at conditions relevant for methanation [15].

Our previous studies focused mainly on the properties of the solid foam, whereas the fluid has only been addressed by ramping the superficial velocity in the laminar regime for air at 1 bar pressure [15,19]. Other CFD studies dealing with heat transport in foams also mainly considered air as fluid with often constant fluid properties such as density, thermal conductivity and specific heat [16,22,31,32]. The study of Razza et al. investigated the influence of wall coupling between foam and tubular wall utilizing He as well as N_2 as fluids [18]. They found significant changes in computed fluid temperatures and justified the deviation through the six times higher thermal conductivity of He (they also assumed constant fluid properties). Additionally, Bianchi et al. [31] found a significant difference for computed heat transfer of either air or water (liquid).

The methanation reaction (CO_2 + $4H_2$ $\rightleftarrows$ CH_4 + $2H_2O$), however, involves gas mixtures (CO_2, H_2 and CH_4), higher velocity ranges (i.e., $\geq$0.5 m s^{-1}) and elevated pressure ranges (4–10 bar) [6]. Furthermore, the effect of gravitational acceleration has not been addressed so far in CFD studies of heat transport in open-cell foams. This might be relevant, because reactors can be operated vertically or horizontally leading to a different effective direction of gravity (i.e., natural convection).

In this study, we thus investigate the influence of the fluid properties on the overall heat transport, heat flows, and solid temperatures; a topic which has largely been neglected in the past. In detail, we analyze the influence of pressure, velocity, fluid composition and gravitational acceleration on the dominant heat transport (i.e., heat removal) mechanisms. We further determine under what circumstances the fluid properties need to be explicitly accounted for to study catalyst carriers for, e.g., the CO_2 methanation reaction.

2. Materials and Methods

The CFD model investigated in this study consists of a foam embedded in a tube (i.e., a tubular structured reactor) and is similar to our previous study [19]. The problem is the steady-state conjugate heat transfer between the flowing fluid and the foam. The simulations were carried out with the commercial CFD software STAR-CCM+ from Siemens PLM (Plano, TX, USA) [33]. Heat production through exothermal reactions is represented by volumetric homogeneously distributed heat sources in the solid ($S = 12.5$ W in all cases) [19]. The heat source intensity of 12.5 W was chosen as it corresponds to the heat released during the CO_2 methanation for a foam in this order of magnitude (50 W for a full sponge equals 12.5 W for one quarter of the sponge) [19].

Two foams (radius 7.5 mm; length 20 mm) are investigated in this study, modeled through Kelvin cell lattices (KC1 and KC2) that differ strongly in their strut diameter (see Figure 1 and Table 1). Only a quarter of the foams as well as the pipe are simulated due to the regularity of the structure and to further reduce computational cost. Polyhedral meshes were created using the integrated STAR-CCM+ meshing utilities (see Figure A1 in Appendix A). The utilized meshes were tested and results were found to be grid independent (see Figure A2 in Appendix B).

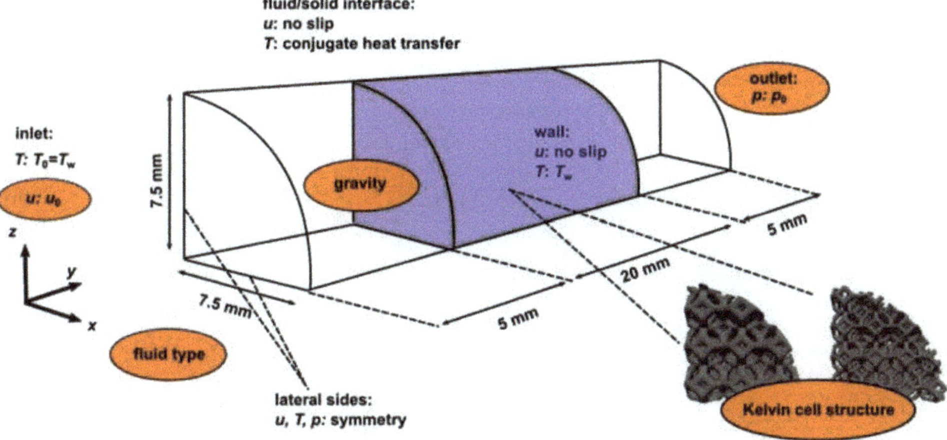

Figure 1. Illustration of model with boundary conditions. The orange highlighted fluid properties are investigated in this study.

Table 1. Properties of the investigated Kell cell lattices.

Parameter, Symbol.	Kelvin Cell 1 (KC1)	Kelvin Cell 2 (KC2)
open porosity, ε_O	0.724	0.845
specific surface area, S_V	1467.8 m^{-1}	1518.9 m^{-1}
cell diameter, d_c	1.924 mm	1.733 mm
strut diameter, d_s	0.591 mm	0.35 mm

The basic equations that are solved for a Newtonian fluid domain are the conservation of mass, momentum and energy:

$$\nabla \cdot (\rho_f \mathbf{U}) = 0, \tag{1}$$

$$\nabla \cdot (\rho_f \mathbf{U} \otimes \mathbf{U}) + \nabla \cdot \left(\mu \left(\nabla \otimes \mathbf{U} + (\nabla \otimes \mathbf{U})^T \right) - \frac{2}{3\mu} (\nabla \cdot \mathbf{U}) \mathbf{I} \right) - \nabla p + \rho_f \mathbf{g} = 0, \tag{2}$$

$$-\nabla \cdot (\rho_f \mathbf{U} h) - \nabla \cdot (\lambda_f \nabla T_f) = 0, \tag{3}$$

with ρ_f denoting the fluid's density, $\mathbf{U}$ the velocity field, h the enthalpy, μ the fluid's dynamic viscosity, T_f the fluid's temperature and λ_f the fluid's thermal conductivity. The solid phase is solely described by the conservation of energy

$$\lambda_s \left(\nabla^2 T_s \right) + S = 0, \tag{4}$$

with λ_s donating the solid thermal conductivity, T_s the solid temperature and S the specific artificial heat source. Furthermore, the fluid density is expressed via the ideal gas law.

The solid properties are kept constant, whereas the fluid properties such as dynamic viscosity, specific heat and thermal conductivity depend on the temperature (see Table 2). The model properties and assumptions are also listed in Table 2 and the basic boundary conditions can be found in Figure 1. In this study, fluid and wall temperatures are kept constant as a fixed temperature boundary at 500 K. This way, the only energy that enters the system is due to the volumetric heat source. Thermal radiation is not explicitly considered in the simulation as the temperatures are moderate and this study wants to investigate heat transport of fluid properties independently. Moreover, the influence of radiation has already been quantified in a comparable setup [15]. Furthermore, turbulence effects were taken into account through a *realizable k-ε RANS turbulence model with All y$^+$ wall-treatment* [34].

Table 2. Properties of this study's model.

Property		Assumption
Fluid dynamic viscosity	μ	Sutherland equation
Fluid heat capacity	$c_{p,f}$	polynomial
Fluid thermal conductivity	λ_f	Sutherland equation
Fluid density	δ_f	ideal gas law
Superficial velocity	v	const. (0.1–4 m s^{-1})
Pore Reynolds number (air)	$Re_p = \dfrac{v \cdot d_s \cdot \rho}{\mu}$	const. (0.3–76)
Wall/inlet temperature	$T_w = T_{in}$	const. (500 K)
Outlet pressure	p	const. (1–10 bar)
Solid heat capacity	$c_{p,s}$	const. (1000 J kg^{-1} K^{-1}).
Solid thermal conductivity	λ_s	const. (5 W m^{-1} K^{-1})
Solid density	δ_s	const. (3950 kg m^{-3})
Solid heat source	S	const. (total: 12.5 W)
Gravitational acceleration		considered
Turbulence		Realizable k-ε RANS (All y$^+$ wall-treatment)
Radiation		neglected [15]

For studying the influence of gravitational acceleration in different directions, gravity is considered in y direction (in flow direction, representing a vertical reactor) as well as in z direction (perpendicular to flow, representing a horizontal reactor). Additionally, the varied and investigated properties are highlighted in orange in Figure 1.

In our previous studies, similar models were validated against pressure drop correlations, heat transfer correlations and verified against CFD data [15,19]. Here, the principial changed model property is the CAD-created geometry. We therefore omit the validation as the models are virtually identical.

3. Results and Discussion

All simulated cases have a total heat source of $S = 12.5$ W, which would correspond to a 50 W total heat source in a full foam setup (i.e., four quarters). As already described, the only thermal energy entering the system is due to the volumetric heat source (i.e., heat production in the solid foam). The global energy balance thus reads

$$S = 12.5 \, \text{W} = Q_{SF} + Q_{SW}, \tag{5}$$

with Q_{SF} being the transferred heat flow from fluid to solid and Q_{SW} being the conducted heat flow from solid to the wall. Which heat removal mechanism dominates (i.e., convection or conduction), can be expressed through the specific heat flow from solid to wall $Q_{SW} S^{-1}$ through non-dimensionalization of Equation (4). For values above 0.5, thermal conduction is the dominant mechanism and for values below 0.5 convection is the dominant one. We note that due to the normalization by the heat source intensity, the actual implemented heat source value becomes less crucial. In our previous study it was shown that in the heat source intensity range between 5 W and 150 W temperature increases (in solid and fluid phases) and heat flows show the almost identical values.

The impact of the several investigated fluid properties on the heat flows is shown in Figure 2. Panel a shows the specific heat flow for a superficial velocity of the two investigated foams up to $4\,\mathrm{m\,s^{-1}}$. Previous studies only investigated heat flows with coupled heat production in the solid has only up to $0.51\,\mathrm{m\,s^{-1}}$ [15,19]. Obviously, both foam structures switch from being dominated by conduction to being dominated by convection at a certain velocity. KC2 switches at a superficial velocity of approximately $v = 1.2\,\mathrm{m\,s^{-1}}$ whereas KC1 switches not until a superficial velocity of approximately $v = 3\,\mathrm{m\,s^{-1}}$. This is most certainly based on the different strut diameters of the foams (KC1: $d_s = 0.591$ mm compared to KC1: $d_s = 0.35$ mm), which ensures a better radial heat removal for KC1 through thermal conduction. This is line with the findings from the studies of Bracconi et al. [11] and Bianchi et al. [21], that investigated the role of the strut diameter in conductive heat transport.

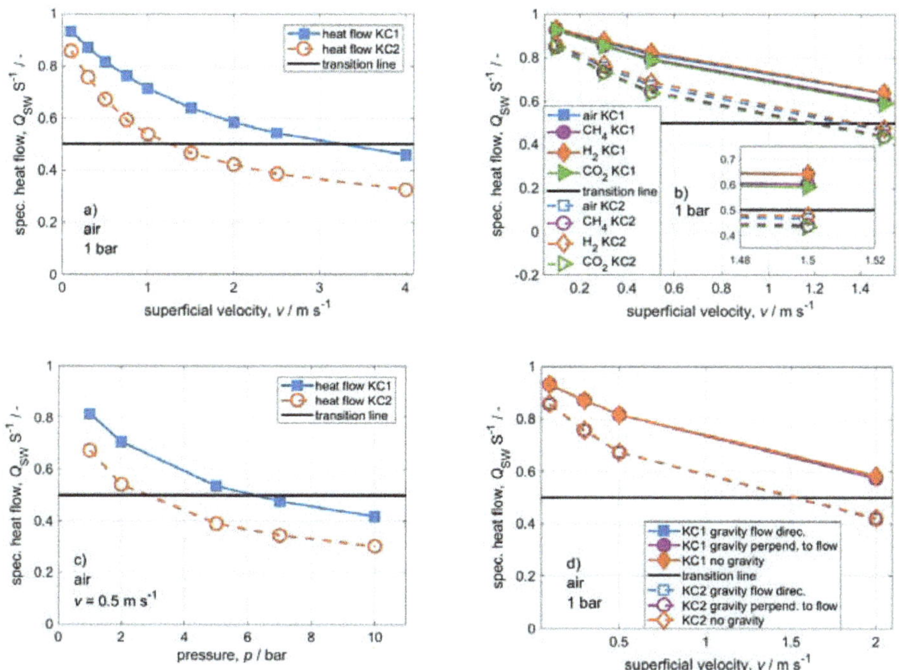

Figure 2. Heat flows for fluid property variation (**a**) Increase in inlet velocity incl. turbulence modelling; (**b**) Influence of fluid type; (**c**) Influence of pressure; (**d**) Influence of gravitational acceleration. Conditions: $S = 12.5\,\mathrm{W}$; $\lambda_s = 5\,\mathrm{W\,m^{-1}\,K^{-1}}$.

Figure 2a shows results using air as fluid as usually done in literature. Panel b shows the impact of the gas type, which is related to the CO_2 methanation, on heat removal. For both KC's, the difference in computed heat flows does not change distinctively, although a small difference between air and hydrogen on the one side and methane and carbon dioxide on the other side becomes apparent. The small differences between the gas types can be explained through the interplay of gas viscosity, density, heat capacity and thermal conductivity which change over temperature for each gas individually. The order of thermal conductivities for the gases at 1 bar from highest to lowest is air, carbon dioxide, methane and hydrogen which is not the same order as in Panel b.

The influence of the pressure on the simulated heat flows at $v = 0.5$ m s^{-1} is shown in Figure 2c. Again, most of the literature investigated heat transport phenomena in foams only at 1 bar absolute pressure [14,35]. With increasing pressure, the heat transport switches from conduction dominated to convection dominated for both foams. At a pressure of 6 bar, both foams dominantly remove heat over convection and hence lose their advantage over standard pellets. The consideration of the actual working pressure thus influences the performance of the catalyst carriers tremendously. Structures that are designed to work efficiently, i.e., conduction dominated, at 1 bar, might already lose their efficiency at 2.5 bar (compare KC2 in panel c). The general effect is obvious, as both gas density and thus the ability of the fluid to cool down the foam via convection increase with pressure.

Often, CFD models omit gravitational acceleration [14,18,19]. Depending on the reactor setup (vertical pipe or horizontal pipe), gravitational acceleration affects the fluid in flow direction (vertical setup) or perpendicular to the flow direction (horizontal setup). Figure 2d shows the specific heat flow as a function of superficial velocity for both foams with gravity in flow direction, perpendicular to the flow as well as neglected gravity. In the investigated superficial velocity range (0.1–2 m s^{-1}), no significant influence of the consideration of gravity could be found. The reason probably lies in forced convection (i.e., pronounced velocity) being a lot more substantial for the overall convection than natural convection (i.e., effect of gravity). To our knowledge, correlations for the Grashof and Rayleigh numbers for this temperature distribution together with the velocity and pressure do not exist. Thus, this result could not have been anticipated by observing dimensionless numbers alone.

The analyzed heat flows for elevated velocities, changing fluid type, influence of pressure and gravity show the influence of the fluid properties on the overall heat transport in open-cell foam carriers. Additionally, Figure 3 shows the corresponding solid temperature rises of the cases shown in Figure 2. For brevity, we only show maximum temperature rises of the solid (instead of additional solid mean and fluid max/mean temperature rises) as they represent the entire solid temperature distribution in foams satisfactory [19]. In Figure 3, it can be seen that the curves of temperature rises of the two structures converge in the convection dominant regime. This behavior happens regardless of the applied pressure or superficial velocity (see panels a and c). Hence, at certain velocities (>3 m s^{-1}) and pressure (>7 bar), the temperature increase seems to be become independent of the foam geometry.

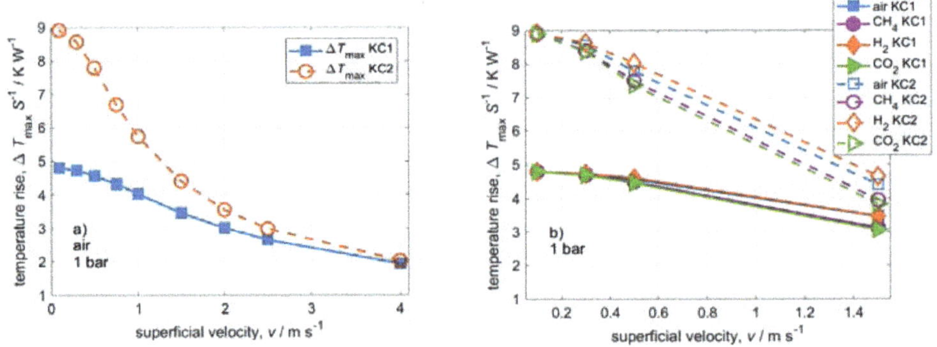

Figure 3. *Cont.*

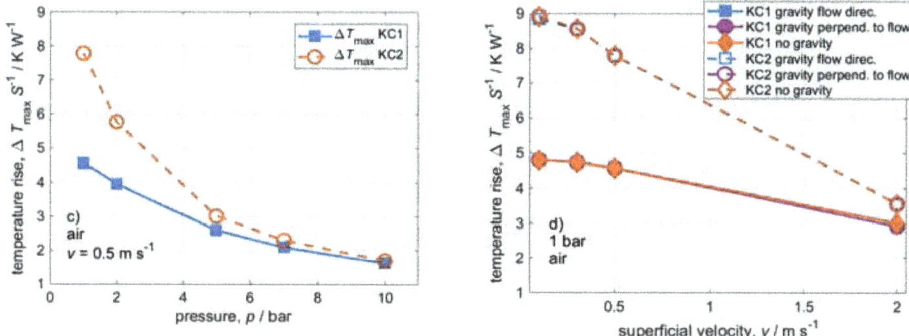

Figure 3. Solid temperature rises per applied heat source intensity for fluid property variation (**a**) Increase in inlet velocity incl. turbulence modelling; (**b**) Influence of fluid type; (**c**) Influence of pressure; (**d**) Influence of gravitational acceleration. Conditions: $S = 12.5$ W; $\lambda_s = 5$ W m^{-1} K^{-1}.

For superficial velocities lower or equal than 1.5 m s^{-1}, the influence of the fluid type on the temperature rise varies for the two structures (see Figure 3b). The effect of the fluid type on the temperature rise increases with convective heat transport. Therefore, KC2 has a more pronounced difference (starting from 0.3 m s^{-1}) than KC1 (difference obvious starting from 1.5 m s^{-1}). This behavior could be expected as the fluid type influences the convective heat transport and becomes more important when the overall contribution of convection increases. To conclude, the fluid type seems to be less influential in the dominant conduction area where a structured foam reactor should be operated. Equivalently to the heat flows, no significant effect of gravity on the temperature rise could be found (Figure 3d).

4. Conclusions

This study analyzed the often-neglected influence of fluid properties on heat transport in open-cell foam reactors. As pointed out, usually heat transport in open-cell foams is investigated using air at 1 bar pressure absolute.

When foams are designed as catalyst carriers (for, e.g., dynamic CO_2 methanation), they should at best remove all heat radially via conduction (here: for working pressure between 4 and 10 bar, the velocity should be lower 0.5 m s^{-1}). At elevated velocities even structures with relatively thick struts can shift into the convection dominated regime. Not only for elevated velocities, but especially for pressure higher than 1 bar (usual operation conditions of CO_2 methanation 4–10 bar) the shift towards convection being dominant might proceed rapidly. In contrast, the choice of fluid mixture seems not as significant as pressure and velocity ranges. Obviously, this has to be checked for other gases of other reactions or inert gases. Nevertheless, air seems a reasonable fluid to study the general heat transport behavior of new foam designs, as the deviations of thermal fields to other gases are not that severe. This way, fluid mixture equations can be omitted and computational effort can be reduced. Additionally, the reactor orientation (represented through gravity) does not seem to have an impact on the heat transport in foams which is relevant for heterogeneous catalysis.

In conclusion, for the design of structured reactors for highly exothermic gas phase reactions, it is crucial to consider the exact velocity and pressure ranges of the investigated reaction. This is especially important, when the catalyst carriers are designed for dynamic or changing load operation. This type of operation might cause tremendous jumps in superficial velocity, pressure or fluid composition which further might influence the overall heat transport and might cause a thermal runaway or serious harm to the catalyst.

Author Contributions: Conceptualization, C.S., J.W., G.R.P. and J.T.; methodology, C.S.; software, C.S.; writing—original draft preparation, C.S.; writing—review and editing, G.R.P. and J.T.; visualization, C.S. and J.W.; supervision, G.R.P. and J.T.; All authors have read and agreed to the published version of the manuscript.

Funding: Christoph Sinn and Jorg Thöming thank the German Research Foundation (DFG) for funding by the priority program SPP 2080 (Katalysatoren und Reaktoren unter dynamischen Betriebsbedingungen für die Energiespeicherung und -wandlung) under grant TH 893/23-1.

Acknowledgments: CS also thanks Harm Ridder for his kind support with the simulation computers that were essential for this study and Kevin Kuhlmann for fruitful discussion about meshing.

Conflicts of Interest: The authors declare no conflict of interest.

Nomenclature

Latin

c_p	isobaric heat capacity, J kg^{-1} K^{-1}
d_c	cell diameter, m
d_s	strut diameter, m
g	gravitational acceleration, 9.81 m s^{-2}
Q	heat flow, W
Q_{SF}	heat flow solid to fluid, W
Q_{SW}	heat flow solid to wall, W
h	specific enthalpy, J kg^{-1}
p	pressure, Pa
S	total heat source intensity, W
S_v	specific surface area, m^{-1}
T	temperature, K
T_w	wall temperature, K
T_{max}	maximum temperature, K
U	velocity, m s^{-1}
v	superficial velocity, m s^{-1}

Greek

ε_O	open porosity, -
μ	dynamic viscosity, Pa s
λ	thermal conductivity, W m^{-1} K^{-1}
ρ	density, kg m^{-3}

Appendix A. Depiction of Volume Meshes

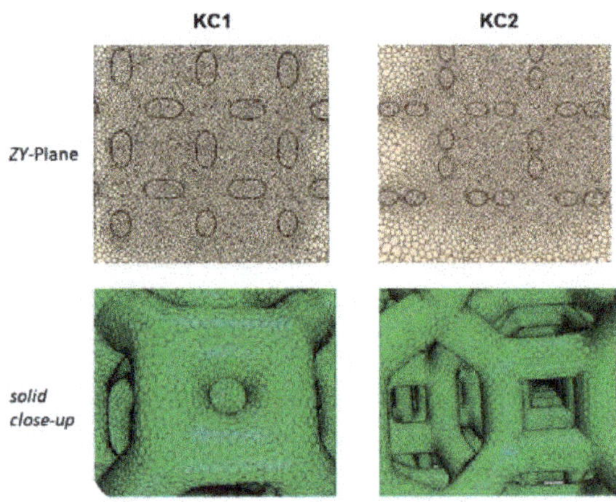

Figure A1. Depiction of volume meshes that were used in this study.

Appendix B. Grid Independence Study

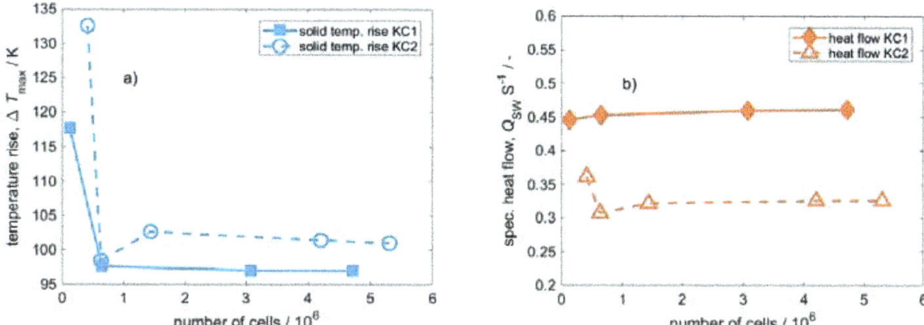

Figure A2. Grid independence study. (**a**) Solid temperature rise from initial 500 K plotted against number of cells; (**b**) specific heat flow from solid to wall plotted against number of cells. For both used geometries, the second-largest meshes (KC1: 3.1 mio. Cells; KC2 4.1 mio cells) were found to give reasonable results. Conditions: air; $p = 1$ bar, $v = 4$ m s^{-1}.

References

1. Kiewidt, L.; Thöming, J. Multiscale modelling of monolithic sponges as catalyst carrier for the methanation of carbon dioxide. *Chem. Eng. Sci. X* **2019**, *2*, 100016. [CrossRef]
2. Stiegler, T.; Meltzer, K.; Tremel, A.; Baldauf, M.; Wasserscheid, P.; Albert, J. Development of a Structured Reactor System for CO$_2$ Methanation under Dynamic Operating Conditions. *Energy Technol.* **2019**, *7*, 1900047. [CrossRef]
3. Gräf, I.; Ladenburger, G.; Kraushaar-Czarnetzki, B. Heat transport in catalytic sponge packings in the presence of an exothermal reaction: Characterization by 2D modeling of experiments. *Chem. Eng. J.* **2016**, *287*, 425–435. [CrossRef]
4. Vogt, C.; Monai, M.; Kramer, G.J.; Weckhuysen, B.M. The renaissance of the Sabatier reaction and its applications on Earth and in space. *Nat. Catal.* **2019**, *2*, 188–197. [CrossRef]
5. Kalz, K.F.; Kraehnert, R.; Dvoyashkin, M.; Dittmeyer, R.; Gläser, R.; Krewer, U.; Reuter, K.; Grunwaldt, J.D. Future Challenges in Heterogeneous Catalysis: Understanding Catalysts under Dynamic Reaction Conditions. *ChemCatChem* **2017**, *9*, 17–29. [CrossRef]
6. Rönsch, S.; Schneider, J.; Matthischke, S.; Schlüter, M.; Götz, M.; Lefebvre, J.; Prabhakaran, P.; Bajohr, S. Review on methanation—From fundamentals to current projects. *Fuel* **2016**, *166*, 276–296. [CrossRef]
7. Kiewidt, L.; Thöming, J. Predicting optimal temperature profiles in single-stage fixed-bed reactors for CO2-methanation. *Chem. Eng. Sci.* **2015**, *132*, 59–71. [CrossRef]
8. Kiewidt, L.; Thöming, J. Pareto-optimal design and assessment of monolithic sponges as catalyst carriers for exothermic reactions. *Chem. Eng. J.* **2019**, *359*, 496–504. [CrossRef]
9. Gräf, I.; Rühl, A.K.; Kraushaar-Czarnetzki, B. Experimental study of heat transport in catalytic sponge packings by monitoring spatial temperature profiles in a cooled-wall reactor. *Chem. Eng. J.* **2014**, *244*, 234–242. [CrossRef]
10. Dixon, A.G.; Partopour, B. Computational Fluid Dynamics for Fixed Bed Reactor Design. *Annu. Rev. Chem. Biomol. Eng.* **2020**, *11*, 109–130. [CrossRef] [PubMed]
11. Bracconi, M.; Ambrosetti, M.; Maestri, M.; Groppi, G.; Tronconi, E. A fundamental analysis of the influence of the geometrical properties on the effective thermal conductivity of open-cell foams. *Chem. Eng. Process. Process Intensif.* **2018**, *129*, 181–189. [CrossRef]
12. Ranut, P.; Nobile, E.; Mancini, L. High resolution X-ray microtomography-based CFD simulation for the characterization of flow permeability and effective thermal conductivity of aluminum metal foams. *Exp. Therm. Fluid Sci.* **2015**, *67*, 30–36. [CrossRef]
13. Iasiello, M.; Bianco, N.; Chiu, W.K.S.; Naso, V. Anisotropic convective heat transfer in open-cell metal foams: Assessment and correlations. *Int. J. Heat Mass Transf.* **2020**, *154*, 119682. [CrossRef]

14. Della Torre, A.; Montenegro, G.; Onorati, A.; Tabor, G. CFD characterization of pressure drop and heat transfer inside porous substrates. *Energy Procedia* **2015**, *81*, 836–845. [CrossRef]

15. Sinn, C.; Kranz, F.; Wentrup, J.; Thöming, J.; Wehinger, G.D.; Pesch, G.R. CFD Simulations of Radiative Heat Transport in Open-Cell Foam Catalytic Reactors. *Catalysts* **2020**, *10*, 716. [CrossRef]

16. Zafari, M.; Panjepour, M.; Emami, M.D.; Meratian, M. Microtomography-based numerical simulation of fluid flow and heat transfer in open cell metal foams. *Appl. Therm. Eng.* **2015**, *80*, 347–354. [CrossRef]

17. Wu, Z.; Caliot, C.; Flamant, G.; Wang, Z. Numerical simulation of convective heat transfer between air flow and ceramic foams to optimise volumetric solar air receiver performances. *Int. J. Heat Mass Transf.* **2011**, *54*, 1527–1537. [CrossRef]

18. Razza, S.; Heidig, T.; Bianchi, E.; Groppi, G.; Schwieger, W.; Tronconi, E.; Freund, H. Heat transfer performance of structured catalytic reactors packed with metal foam supports: Influence of wall coupling. *Catal. Today* **2016**, *273*, 187–195. [CrossRef]

19. Sinn, C.; Pesch, G.R.; Thöming, J.; Kiewidt, L. Coupled conjugate heat transfer and heat production in open-cell ceramic foams investigated using CFD. *Int. J. Heat Mass Transf.* **2019**, *139*, 600–612. [CrossRef]

20. Ambrosio, G.; Bianco, N.; Chiu, W.K.S.; Iasiello, M.; Naso, V.; Oliviero, M. The effect of open-cell metal foams strut shape on convection heat transfer and pressure drop. *Appl. Therm. Eng.* **2016**, *103*, 333–343. [CrossRef]

21. Bianchi, E.; Schwieger, W.; Freund, H. Assessment of Periodic Open Cellular Structures for Enhanced Heat Conduction in Catalytic Fixed-Bed Reactors. *Adv. Eng. Mater.* **2016**, *18*, 608–614. [CrossRef]

22. Diani, A.; Bodla, K.K.; Rossetto, L.; Garimella, S.V. Numerical investigation of pressure drop and heat transfer through reconstructed metal foams and comparison against experiments. *Int. J. Heat Mass Transf.* **2015**, *88*, 508–515. [CrossRef]

23. Du, S.; Tong, Z.X.; Zhang, H.H.; He, Y.L. Tomography-based determination of Nusselt number correlation for the porous volumetric solar receiver with different geometrical parameters. *Renew. Energy* **2019**, *135*, 711–718. [CrossRef]

24. Aguirre, A.; Chandra, V.; Peters, E.A.J.F.; Kuipers, J.A.M.; Neira D'Angelo, M.F. Open-cell foams as catalysts support: A systematic analysis of the mass transfer limitations. *Chem. Eng. J.* **2020**, *393*, 124656. [CrossRef]

25. Dong, Y.; Korup, O.; Gerdts, J.; Roldán Cuenya, B.; Horn, R. Microtomography-based CFD modeling of a fixed-bed reactor with an open-cell foam monolith and experimental verification by reactor profile measurements. *Chem. Eng. J.* **2018**, *353*, 176–188. [CrossRef]

26. Della Torre, A.; Lucci, F.; Montenegro, G.; Onorati, A.; Dimopoulos Eggenschwiler, P.; Tronconi, E.; Groppi, G. CFD modeling of catalytic reactions in open-cell foam substrates. *Comput. Chem. Eng.* **2016**, *92*, 55–63. [CrossRef]

27. Wehinger, G.D.; Heitmann, H.; Kraume, M. An artificial structure modeler for 3D CFD simulations of catalytic foams. *Chem. Eng. J.* **2016**, *284*, 543–556. [CrossRef]

28. Bracconi, M.; Ambrosetti, M.; Maestri, M.; Groppi, G.; Tronconi, E. A fundamental investigation of gas/solid mass transfer in open-cell foams using a combined experimental and CFD approach. *Chem. Eng. J.* **2018**, *352*, 558–571. [CrossRef]

29. Frey, M.; Bengaouer, A.; Geffraye, G.; Edouard, D.; Roger, A.C. Aluminum Open Cell Foams as Efficient Supports for Carbon Dioxide Methanation Catalysts: Pilot-Scale Reaction Results. *Energy Technol.* **2017**, *5*, 2078–2085. [CrossRef]

30. Montenegro Camacho, Y.S.; Bensaid, S.; Lorentzou, S.; Vlachos, N.; Pantoleontos, G.; Konstandopoulos, A.; Luneau, M.; Meunier, F.C.; Guilhaume, N.; Schuurman, Y.; et al. Development of a robust and efficient biogas processor for hydrogen production. Part 2: Experimental campaign. *Int. J. Hydrogen Energy* **2018**, *43*, 161–177. [CrossRef]

31. Bianchi, E.; Groppi, G.; Schwieger, W.; Tronconi, E.; Freund, H. Numerical simulation of heat transfer in the near-wall region of tubular reactors packed with metal open-cell foams. *Chem. Eng. J.* **2015**, *264*, 268–279. [CrossRef]

32. Nie, Z.; Lin, Y.; Tong, Q. Modeling structures of open cell foams. *Comput. Mater. Sci.* **2017**, *131*, 160–169. [CrossRef]

33. STAR-CCM+. Available online: https://www.plm.automation.siemens.com/global/en/products/simcenter/STAR-CCM.html (accessed on 6 October 2020).

34.   Wehinger, G.D. Radiation Matters in Fixed-Bed CFD Simulations. *Chemie-Ingenieur-Technik* **2019**, *91*, 583–591. [CrossRef]
35.   Iasiello, M.; Cunsolo, S.; Bianco, N.; Chiu, W.K.S.; Naso, V. Developing thermal flow in open-cell foams. *Int. J. Therm. Sci.* **2017**, *111*, 129–137. [CrossRef]

 *chemengineering*

Article

3-D Multi-Tubular Reactor Model Development for the Oxidative Dehydrogenation of Butene to 1,3-Butadiene

Jiyoung Moon, Dela Quarme Gbadago and Sungwon Hwang *

Graduate School of Chemical and Chemical Engineering, Inha University, 100 Inha-ro, Michuhol-gu, Incheon 22212, Korea; jiyoung.moon@inha.edu (J.M.); delaquarme@inha.edu (D.Q.G.)
* Correspondence: sungwon.hwang@inha.ac.kr; Tel.: +82-32-860-7461

Received: 29 April 2020; Accepted: 16 July 2020; Published: 21 July 2020

Abstract: The oxidative dehydrogenation (ODH) of butene has been recently developed as a viable alternative for the synthesis of 1,3-butadiene due to its advantages over other conventional methods. Various catalytic reactors for this process have been previously studied, albeit with a focus on lab-scale design. In this study, a multi-tubular reactor model for the butadiene synthesis via ODH of butene was developed using computational fluid dynamics (CFD). For this, the 3D multi-tubular model, which combines complex reaction kinetics with a shell-side coolant fluid over a series of individual reactor tubes, was generated using OpenFOAM®. Then the developed model was validated and analyzed with the experimental results, which gave a maximum error of 7.5%. Finally, parametric studies were conducted to evaluate the effect of thermodynamic conditions (isothermal, non-isothermal and adiabatic), feed temperature, and gas velocity on reactor performance. The results showed the formation of a hotspot at the reactor exit, which necessitates an efficient temperature control at that section of the reactor. It was also found that as the temperature increased, the conversion and yield increased whilst the selectivity decreased. The converse was found for increasing velocities.

Keywords: computational fluid dynamics (CFD); multi-tubular reactor; oxidative dehydrogenation (ODH); reactor design; butadiene; multiscale modeling

---

1. Introduction

Hydrocarbons such as butene and 1,3-butadiene (butadiene) are important building block chemicals [1]. For example, butadiene is a major product of the petrochemical industry, largely used in the production of synthetic rubbers such as styrene-butadiene-rubber (SBR) and polybutadiene rubber (PBR), which are intermediates used to produce tires and plastic materials. The majority of butadiene, which accounts for over 95% of global production, has been produced as a byproduct of the steam cracking of hydrocarbons such as naphtha [2,3]. To date, steam crackers are the main process units to produce butadiene. In this process, the mixture of steam and hydrocarbon is used as feedstock and heated to extremely high temperatures [1]. Then naphtha, which is composed of heavier hydrocarbons (C5–C12), breaks into lighter hydrocarbons such as ethylene, propylene, and other light olefins [4]. Subsequently, a separation process is required after steam cracking, and butadiene is extracted by an aprotic polar solvent from crude C4 raffinates [5]. However, this conventional route is not only an energy-consuming process because the reaction is highly endothermic, but also requires additional energy to activate the reactant [6]. The reaction at high temperature also leads to the deposition of coke formed by the pyrolysis of hydrocarbon with reported high CO_2 emissions [7]. In recent years, using natural gas and refinery gas as feedstock has increased and the production of naphtha-based ethylene has also decreased due to the shale gas revolution. As a result, only a small amount of butadiene can be obtained to meet the growing global demand, which leads to a shortage of butadiene supply [5,8]. Thus, a more efficient butadiene production process is required in terms of both economic and environmental perspectives. Consequently, alternative processes such as direct dehydrogenation of n-butane and oxidative dehydrogenation of n-butene have been widely studied [6,9]. Catalytic or direct dehydrogenation reactions occur on metal surfaces and acid sites where alkene (or alkane) decomposes into olefin and hydrogen [7,8]. Although there are several commercial catalytic dehydrogenation processes including Catofin and UOP Oleflex [10], a few problems that inhibit the commercialization of the direct dehydrogenation have been found. First of all, the process is highly endothermic, so the reaction occurs at a relatively high temperature (e.g., 500 to 650 °C) [9,11]. Furthermore, such severe reaction conditions cause side reactions such as thermal cracking and rapid coke formation on the catalysts, which leads to catalyst deactivation [12]. To overcome such limitations, oxidative dehydrogenation (ODH) of n-butene has been proposed and has been widely studied in recent years [9–11]. ODH is an exothermic reaction at relatively low temperature (e.g., 300 to 400 °C), which takes place in the presence of oxygen on the surface of catalysts, producing C4 olefin. Thus, the exothermic reaction at a lower temperature can reduce a substantial amount of energy compared to the direct dehydrogenation process [7]. Furthermore, the presence of oxygen easily prevents coke deposition on the catalyst surface through promoting coke combustion. However, some challenging issues are still presented as the selectivity of the product is strongly dependent on the reaction condition used [7,13]. Multiple byproducts are produced by the ODH reaction, and the gas phase oxygen in lattice and butene (or butane) can also form combustion products such as carbon dioxide and water. Hence, previous studies have been extensively conducted to improve the selectivity and yield of butadiene at mild reaction conditions to enable the ODH process to compete with steam cracking technology [7].

For this, researchers have developed novel and efficient catalysts for the ODH process. The catalysts that consist of unsupported metal oxides containing two or more transition metal components such as bismuth molybdates (Bi–Mo) or Bi–Mo combined with V, Fe, Zr, and ferrites, have been extensively used for ODH reaction. V/MgO composites and Cr_2O_3/Al_2O_3 catalysts are also used [14,15]. Conversely, Park et al. [11] suggested the use of metal-free carbon nanomaterials as potential catalysts due to their environmental benignity and unique surface characteristics. In particular, ferrite and Bi–Mo catalysts have been extensively used due to their high catalytic activity and selectivity [7,16]. Typically, reaction mechanisms and kinetics are identified before modeling the reaction system for reactor design [17]. Several mechanisms have been suggested for this catalytic reaction. Sterrett and McIlvried [18] derived semi-empirical kinetic equations for the ODH of butenes to butadiene over a zinc–chromium–ferrite catalyst in a fixed bed reactor, which appears to be an attractive and flexible continuous process for producing butadiene. Ding et al. [19] also investigated the ODH kinetics over ferrite catalysts based on the Mars–Van Krevelen mechanism and proposed redox model rate equations. Téllez et al. [20] carried out a kinetic study of the ODH of butane over V/MgO catalysts and statically analyzed several competing kinetic models such as the Langmuir–Hinshelwood or Mars–Van Krevelen type by extensive experimental work. The study showed that the Mars–Van Krevelen mechanism best fits the reaction data. Recently, kinetic models for novel catalysts such as VO_x/Al_2O_3, $CrOxVOx/MCM$-41, and (Ni, Fe, Co)–Bi–O/γAl_2O_3 have also been developed [21–24].

With the development of kinetic models, efforts have been made to design optimal reactors for the ODH reaction [25]. Several types of reactors have been proposed for this reaction and the adiabatic fixed-bed reactor is known for its simplest design and good product yield and selectivity [26]. The adiabatic operation is important in terms of commercialization because of its conventional usage [18]. However, such systems cannot control the temperature inside the reactor, resulting in more side products [25]. On the other hand, in a fluidized reactor, the uniform particle mixing is achieved and can maintain the temperature and product uniformly across the reactor. Therefore, it has been recommended for highly exothermic and temperature-sensitive reactions. However, the complex behavior of the fluid bed and high initial capital cost due to the expansion and circulation of solid materials in the reactor are known as major drawbacks [27]. A multi-tubular reactor, which is commonly used for several industrial processes, has shown good temperature control for highly exothermic reactions by removing heat using a shell side coolant. The coolant either flows co-currently or counter-currently to achieve the isothermal condition in the reactor because it is essential to avoid rapid temperature increase to maintain the product yield and selectivity [25]. Xingan et al. [16] compared the three above-mentioned reactors on a laboratory and pilot-scale for the ODH of butene to butadiene and indicated that the multi-tubular reactor showed the highest efficiency. Even though the regeneration or replacement of catalysts after deactivation presented challenges for this reactor, oxygen in the feed gas promoted the coke combustion in the ODH reaction, resulting in low catalyst deactivation. For this reason, it was found that the multi-tubular reactor with coolant in the shell-side is appropriate for the ODH reaction.

Particularly in the modeling of reaction systems for reactor design, computational fluid dynamics (CFD) is a well-established tool for predicting the accurate fluid patterns and understanding the transport phenomena and its influences on reactions in chemical engineering. It has also shown great advantages in evaluating and validating the performance of reactors before their actual fabrication and scale up to a pilot plant [25,28]. Therefore, many models have been built for the design and optimization of various types and configurations of reactors using CFD. For instance, Huang et al. [29] developed a two-dimensional multi-scale model for the ODH of butene in the fixed-bed reactor. A more realistic description of flow within a fixed bed reactor was investigated by Dixon et al. [30] and a comprehensive 2D model was also suggested to simulate the flow behavior and reaction by Chen et al. [31]. With respect to multi-tubular reactor modeling, CFD modeling coupled with a reaction model was investigated to predict the effect of coolant flow rate and baffle configurations on *o*-xylene conversion [28]. Additionally, Fattahi et al. [32] examined a variety of parameters affecting the performance of a multi-tubular system for the ODH of ethane. Furthermore, in the previous work of Mendoza et al. [25], an isothermal multi-tubular fixed-bed reactor was developed for ODH reaction; however, it was found that there were a few limitations. Considering that the model was reduced to a single tubular reactor for simplification, the description of the model such as the flow pattern of the shell-side coolant and the actual transfer phenomena in the reactor could not be represented sufficiently. In addition, the reaction kinetics were assumed to be simple pseudo-first-order kinetics derived from the equations by Ding et al. [19], which showed discrepancies toward the experimental data.

Therefore, this study aimed to develop the 3D multi-tubular reactor model to accurately describe the fluid flow and predict the reactor performance by using CFD simulation. Due to the multi-physics capability of CFD models, complex geometric engineering problems can be easily modeled and simulated. Hence, to provide useful insights into the operation of industrial shell and tube type reactors, the CFD simulation was executed. Rigorous mathematical modeling of the actual reaction kinetics over a powder Zn–Cr–Fe catalyst as reported in Sterrett and McIlvried [18] was considered. The developed model represents the industrial application of butadiene synthesis to provide insight into the behavior of this process by analyzing the influence of key operating conditions such as flow rates and temperature on the reactor performance, which was not considered in any of the review literature. Moreover, due to the high exothermicity of the reaction process, temperature control was implemented using water as the shell side coolant. The mathematical formulation of the model in terms of mass continuity, momentum transport, and reaction kinetics is presented in Section 2. Section 3 comprises the CFD simulation methods such as geometry and mesh development, design and operating parameters, and solution strategy. The validation and analysis of the results is presented in the results and discussion session (Section 4). The study finally concludes with the findings in Section 5.

2. Model Development

The present study represented the gas phase exothermic reaction of butadiene synthesis via ODH reaction over Zn–Fe–Cr catalysts where the shell side coolant counter-currently flows to maintain the isothermal condition of tubes. The catalyst used in this study had an approximately 1:1:1 proportion by mass of zinc, chromium, and iron, therefore, the catalyst bulk density was estimated at 7393.92 kg/m³. With the porosity of 0.35, the catalyst density was calculated to be 4806.05 kg/m³. [18]. The porous zone conditions were specified to reflect the pressure drop due to porous resistance along the catalyst bed.

2.1. Mass Conservation

The equation for conservation of mass, or continuity equation, can be written as follows:

$$\frac{\partial \varrho}{\partial t} + \nabla \cdot (\varrho \vec{u}) = R \tag{1}$$

Equation (1) is valid for incompressible as well as compressible flows and the source term, R, is defined as the mass source term due to chemical reaction.

In this system, three gas-phase reactions occurred and the overall species transport equation can be written as:

$$\frac{\partial}{\partial t}(\varrho Y_i) + \nabla \cdot (\varrho \vec{u} Y_i) - \nabla \cdot (\varrho D_i \nabla Y_i) = m_i + R_i \tag{2}$$

where Y_i is the mass fraction of species i, and Di is the mass diffusion coefficient of species i in the mixture. Equation (2) is solved for $N - 1$ species where N is the total number of chemical species present in the system with the total mass fraction of species in the mixture equal to unity (Equation (3)).

$$\sum_{i=1}^{n} Y_i = 1 \tag{3}$$

2.2. Momentum Conservation

The momentum equation is described as:

$$\frac{\partial}{\partial t}(\rho \vec{u}) + \nabla \cdot (\rho \vec{u} \vec{u}) = \nabla \cdot (\mu \nabla \rho \vec{u}) - \nabla p + \rho \vec{g} + \dot{m} \vec{u} + \vec{S} \tag{4}$$

where μ is the viscosity; p is the pressure; $\vec{g}$ is the gravitational acceleration vector; and $\dot{m}$ is the contribution of mass transfer. The porous media is applied to determine the pressure loss in the flow through the catalytic packed beds, and modeled by the addition of a momentum source term $\vec{S}$ in the flow equation (Equation (4)). The flow resistance defined by Equation (5) is written as:

$$\vec{S} = \frac{\mu}{\alpha} \vec{u} + C \frac{1}{2} \rho |\vec{u}| \vec{u} \tag{5}$$

The source term $\vec{S}$, composed of two terms: the viscous and inertial terms in that order, where α is the permeability of the porous media and C is the inertial resistance.

Turbulent Modeling

Due to the diameter of the tubes and density of the gases, the tube side Reynolds number was calculated to be 2428 in the laminar region. As such, the tubes were modeled as a laminar flow.

With respect to the shell side coolant, the flow was assumed to be turbulent due to the density, velocity, and the viscosity of coolant. Hence, the standard k-epsilon turbulent model by Launder and Spalding [33] was adopted for this study. This model has gained in popularity due to its robustness and reasonable accuracy for a wide range of turbulent flows and valid for fully turbulent flows. It is a semi-empirical model consisting of transport model equations for the turbulence kinetic energy (k) and its dissipation rate (ε). The turbulent viscosity is written as:

$$\frac{\partial}{\partial t}(\rho k) = \nabla \cdot \left(\rho \frac{\mu_t}{\sigma_k} \nabla k \right) + G_k + \frac{2}{3} \rho (\nabla \cdot \vec{u}) k - \rho \varepsilon + S_k \tag{6}$$

In Equation (6), G_k is the turbulence kinetic energy generated by the mean velocity gradients; σ_k is an empirical constant; S_k is the source term; and μ_t is the turbulent viscosity defined by Equation (7) as follows:

$$\mu_t = C_\mu \frac{k^2}{\varepsilon} \tag{7}$$

where C_μ is the Eddy-viscosity coefficient and the dissipation rate of k is obtained from Equation (8).

$$\frac{\partial}{\partial t}(\rho\varepsilon) = \nabla\cdot\left(\rho\frac{\mu_t}{\sigma_\varepsilon}\nabla\varepsilon\right) + \frac{C_1 G_k \varepsilon}{k} - \frac{2}{3}C_1\rho(\nabla\cdot\vec{u})\varepsilon - C_2\rho\frac{\varepsilon^2}{k} + S_\varepsilon \tag{8}$$

C_1, C_2, and σ_ε are empirical constants and values for constants used in this study were determined as follows [33]: $C_\mu = 0.09$, $\sigma_k = 1.0$, $\sigma_\varepsilon = 1.3$, $C_1 = 1.44$, $C_2 = 1.92$.

2.3. Reaction Kinetics

The oxidative dehydrogenation of butene to butadiene reactions occur in the gas phase with butadiene, carbon dioxide, and water as the only reactor products; according to the experimental study of [18] using a zinc–chromium–ferrite catalyst. The reaction equations (Equations (9)–(11)) are as follows:

$$C_4H_8 + \frac{1}{2}O_2 \rightarrow C_4H_6 + H_2O \tag{9}$$

$$C_4H_8 + 6O_2 \rightarrow 4CO_2 + 4H_2O \tag{10}$$

$$C_4H_6 + \frac{11}{2}O_2 \rightarrow 4CO_2 + 3H_2O \tag{11}$$

Sterrett et al. [18] developed the rate of formation of butadiene and carbon dioxide over the range of conditions explored in the kinetic runs. The carbon dioxide production data fitted well with zero-order kinetics, although the rate model that gave the best fit was found to be the semi-empirical and two-site model. Equation (12) indicates the butadiene formation from butene and the butadiene consumption by combustion is represented in Equation (13). Equation (14) is the rate of carbon dioxide formation resulting from the combustion of both butene and butadiene. From Equations (12)–(14), the rate of reactions of all species can be expressed by considering the stoichiometric coefficients of each reaction equations (Equations (15)–(18)). The kinetic parameters used here are presented in Table 1 below.

$$r_{BD1} = \frac{A_1 e^{-E_1/RT} p_B{}^n p_{O_2}}{(1 + K_{O_2}p_{O_2})(1 + K_B p_B + K_{BD}p_{BD})} \tag{12}$$

$$r_{BD2} = \frac{A_2 e^{-E_2/RT} p_{O_2} p_{BD}}{(1 + K_{O_2}' p_{O_2})(p_{BD} + p_B)} \tag{13}$$

$$r_{CO_2} = \frac{A_2 e^{-E_2/RT} p_{O_2}}{1 + K_{O_2}' p_{O_2}} \tag{14}$$

$$r_{BD} = r_{BD1} - r_{BD2} \tag{15}$$

$$r_B = -r_{BD1} - \frac{1}{4}(r_{CO_2} - 4r_{BD2}) \tag{16}$$

$$r_{O_2} = -0.5r_{BD1} - \frac{6}{4}(r_{CO_2} - 4r_{BD2}) - \frac{11}{2}r_{BD2} \tag{17}$$

$$r_{H_2O} = r_{BD1} + (r_{CO_2} - 4r_{BD2}) + 3r_{BD2} \tag{18}$$

Table 1. Kinetic parameters of reaction rates for butadiene and carbon dioxide production.

Parameter	Unit	Value
A_1	kmol/kgcat-s	1.66×10^9
A_2	kmol/kgcat-s	6.69×10^3
E_1	J/mol	139745.6
E_2	J/mol	87027.2
K_B	atm^{-1}	0
K_{BD}	atm^{-1}	285
K_{O_2}	atm^{-1}	415
K_{O_2}	atm^{-1}	1800
n	-	0.142

2.4. Energy Equations

The energy conservation equation is described as:

$$\frac{\partial}{\partial t}(\rho(H+K)) + \nabla(\rho\vec{u}(H+K)) - \nabla\cdot(\alpha^{-\text{eff}}\nabla H) = \frac{\partial \bar{p}}{\partial t} + \rho\vec{g}\cdot\vec{u} + \dot{Q}_R \tag{19}$$

where H is the enthalpy of species; K is the kinetic energy; $\alpha^{-\text{eff}}$ is the effective thermal diffusivity; and $\dot{Q}_R$ is the rate of heat change due to reactions.

3. Computational Fluid Dynamics (CFD) Simulation Methods

3.1. Geometry and Mesh Generation

3.1.1. Geometric Description of the Multi-Tubular Reactor

In this study, a 3D shell and tube type multi-tubular reactor was designed by SALOME®, an open-source CAD software, as shown in Figure 1. In the previous work by Mendoza et al. [25], a multi-tubular reactor model was reduced to a single tubular reactor and the effect of reactor diameter and types of coolants were analyzed. However, this study presents a shell and tube type reactor close to the actual industrial application to accurately investigate the reactor performance. The reactor has both shell and tube side, and the reaction takes place at tube sides while the coolant flows along the shell side counter-currently.

The model is a single pass shell and tube reactor on both the tube and shell side. The Tubular Exchanger Manufacturers Association (TEMA) [34] standards were used as the design approach for a 14 tubular reactor with a triangular pitch. A triangular pattern was adopted because it is possible to arrange more tubes and enhance heat transfer efficiency. The tube has a length of 81.28 cm (32 in) and an inside diameter of 2.54 cm (1 in) filled with the solid Zn–Fe–Cr powder catalyst [18], with a pitch of 3.18 cm, a shell side diameter and length of 21.33 cm and 81.28 cm, respectively. Baffles were introduced to increase the heat transfer rate for efficient temperature control. These are conventional segmental baffles with a baffle cut of 42% and baffle spacing of 20.32 cm. These and other geometric dimensions of the multi-tubular reactor are shown in Table 2.

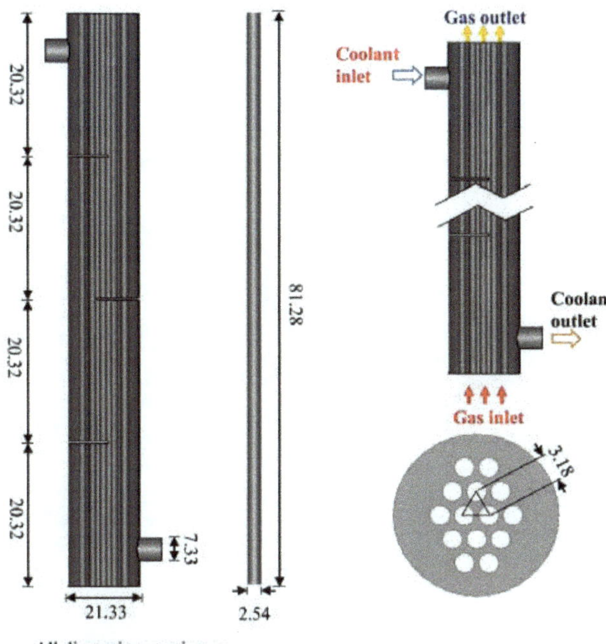

All dimensions are in cm
Note: not drawn to scale

Figure 1. 3D multi-tubular reactor model.

Table 2. Geometric dimensions of the multi-tubular reactor.

Parameters	Unit	Value
Shell diameter	cm	21.33
Shell length	cm	81.28
Tube diameter	cm	2.54
Tube length	cm	81.28
Tube pitch	cm	3.18
Tube pattern	-	30°-triangular
Number of tubes	-	14
Nozzle diameter	cm	7.33
Nozzle length	cm	5.6
Baffle spacing	cm	20.32
Baffle thickness	cm	0.37
Number of baffles	-	3

3.1.2. Mesh Generation

Non-conformal hexahedral dominant meshes were generated for the CFD simulation. The tube side consisted of only hexahedral meshes to enhance the stability of the numerical simulation while the shell side comprised tetrahedral meshes, as shown in Figure 2.

Complex geometries come with high computational cost and resources and as such, a good compromise between mesh size and accuracy needs to be made. A preliminary study using three different mesh sizes was conducted to find the optimal mesh size for the CFD model. Mesh numbers of 1,387,210 (Mesh A), 757,921 (Mesh B), and 532,192 (Mesh C) were generated and investigated using the mole fractions, density, and velocity profile (Section 4.1).

The tube side mesh was generated with OpenFOAM's in-built *SnappyHexMesh* utility tool with the maximum orthogonal quality of 27.99 and skewness of 0.64 and an aspect ratio of 3.38. The shell side mesh, on the other hand, was generated with SALOME® CADsd software. The two meshes were then merged together using OpenFOAM® *mergeMesh* utility. This technique of generating two independent meshes before merging them together allows for a reduction in the computational domain (number of meshes) without compromising the accuracy and convergence of the numerical methods.

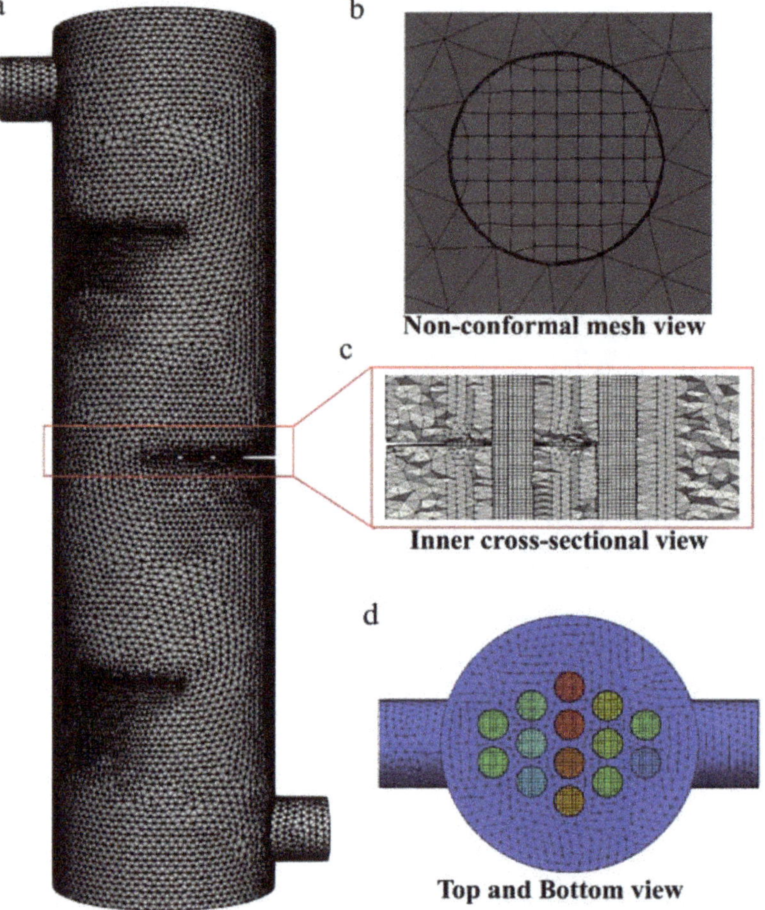

Figure 2. Generated multi-tubular reactor mesh for the CFD simulation. (**a**) External view of the reactor; (**b**) Non-conformal mesh view; (**c**) Detailed cross-sectional view; (**d**) Bottom and top view showing the 14 tubes.

3.2. Model Parameters Used in the Computational Fluid Dynamics (CFD) Model

Design and model parameters such as operating conditions, catalyst properties, and boundary conditions are provided in Table 3. The operating conditions and design parameters were obtained from the experimental work of Sterrett et al. [18]. To validate the developed CFD model, a set of experimental data was chosen for comparison. In the experimental study, the inlet gas consisting of butene mixed with nitrogen, water, and oxygen was fed into the fixed bed reactor at 633 K where the outlet partial pressure values were measured for five different weight hourly space velocities (WHSV).

As the reactions are exothermic, steam and nitrogen were fed in the reactor as heat sinks for the purpose of maintaining the isothermal conditions. The gas velocity used in the CFD simulation was calculated based on the WHSV. Water was selected as a shell side coolant due to its high specific heat capacity and injected at room temperature (300 K) to enhance temperature control and ensure isothermality of the reactor tubes.

Table 3. Process conditions and model parameters.

Parameters	Unit	Value
Operating pressure	atm	1
Gas inlet temperature	K	633
Gas velocity	m/s	variant
Coolant inlet temperature	K	300
Coolant velocity	m/s	1
Inlet mole compositions		
Butene (C_4H_8)	-	0.05
Oxygen (O_2)	-	0.047
Steam (H_2O)	-	0.717
Nitrogen (N_2)	-	0.186
Butadiene (C_4H_6)	-	0
Carbon dioxide (CO_2)	-	0
Bulk density of catalyst	kg/m^3	7393.92
Porosity	-	0.35
Actual catalyst density	kg/m^3	4806.05
k	m^2/s^2	0.00375
ε	m^2/s^3	0.012

3.3. Solution Strategy

The CFD simulation was executed with OpenFOAM®, an opensource tool for computational fluid dynamics. To include the effect of mass and momentum sources, the equation of continuity and momentum in the adopted OpenFOAM® solver (*chtMultiRegionFoam*) were modified. The PIMPLE algorithm, a dynamic solver, was applied to solve for pressure-velocity coupling. This allows for the simulation of flows with large time steps in complex geometries such as multi-tubular reactors. For the stability, convergence, and accuracy of the numerical solution, the time step was adjusted after every complete iteration loop by fixing the Courant number to 1. The Courant number measures the traversal of information across the cells in the mesh domain and is recommended to be less than 1. The overall solution strategy performed in the present study is shown in Figure 3. The developed model was simulated by a computer with an Intel Xeon® processor (2.10 GHz, 16 cores) and 64 GB of RAM. For each simulation, the computational time was 23 h, which was a total of 230 h for the ten cases.

A gas-phase catalytic reaction simulation through a particle-filled fixed bed reactor was executed. The detailed description of numerical discretization techniques and boundary conditions in this work adopted from our previous study [35] are given in Tables 4 and 5. The packed bed was assumed as a continuous porous media to include the effect of pressure drop and heat transfer. Therefore, the porous medium flows can be incorporated by adding viscous and inertial resistance terms in the momentum equations. Consequently, the results obtained by simulation were post-processed with Paraview®, an open-source graphical Virtualization Tool Kit (VTK).

Table 4. Computational boundary conditions.

Boundary field	Gas inlet	Coolant inlet	Gas outlet	Coolant outlet	Wall
Gas velocity	*fixedValue*	*fixedValue*	*zeroGradient*	*zeroGradient*	*noSlip*
Coolant velocity	*fixedValue*	*fixedValue*	*zeroGradient*	*zeroGradient*	*noSlip*
Pressure	*fixedFlux Pressure*	*fixedFlux Pressure*	*totalPressure*	*totalPressure*	*fixedFlux Pressure*
Temperature	*fixedValue*	*fixedValue*	*inletOutlet*	*inletOutlet*	*zeroGradient*
Species concentration	*fixedValue*	*fixedValue*	*inletOutlet*	*inletOutlet*	*zeroGradient*
Epsilon	*fixedValue*	*fixedValue*	*zeroGradient*	*zeroGradient*	*epsilonWall Function*
K	*fixedValue*	*fixedValue*	*zeroGradient*	*zeroGradient*	*kqRWall Function*

The italicized words indicate default or standard boundary condition names in OpenFOAM.

Table 5. Numerical discretization schemes.

Term	Keyword	Description	Scheme
Convection	*divSchemes*	Discretizing the divergence, ∇	*Gauss limitedLinear V 1*
Gradient	*gradSchemes*	Discretizing the gradient, ∇	*Gauss linear 1*
Diffusion	*laplacianSchemes*	Discretizing the Laplacian, ∇^2	*Gauss linear orthogonal*
Time derivative	*ddtSchemes*	Discretizing first and second-order term derivatives, $\frac{\partial}{\partial t}$, $\frac{\partial^2}{\partial t^2}$	*Euler*
Others	*interpolationSchemes* *snGradSchemes*	Cell to face interpolations Component of gradient normal to a cell face	*linear orthogonal*

The italicized words indicate default or standard numerical scheme names in OpenFOAM.

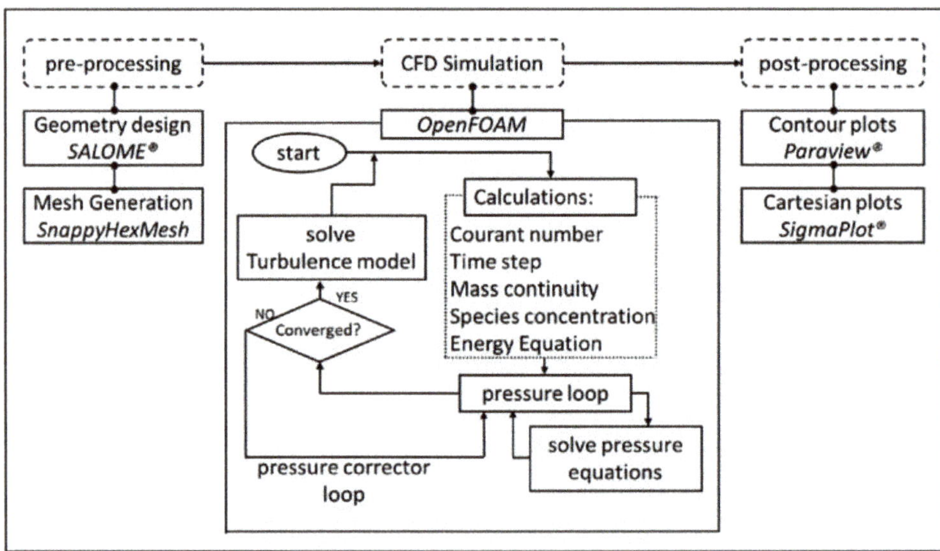

Figure 3. Summary of the computational fluid dynamics (CFD) solution strategy.

4. Results and Discussion

This chapter comprises two sections. In the first section, the developed 3D multi-tubular reactor model was validated with the experimental results of Sterrett et al. [18] and evaluated under both adiabatic and isothermal conditions. In the second section, a parametric study on the reactor performance was carried out under different thermodynamic conditions and by varying gas velocity and temperature. In each study, gas velocity was varied from 3.25 m/s to 35.31 m/s and the temperature from 610 K to 690 K.

4.1. Mesh Independence Test

Figure 4 shows the mesh independence test results for three different mesh sizes in terms of mole fractions, density, and gas velocity. Meshes A and B predicted the variables to the same extent with only a little deviation in the density predictions. However, mesh C overpredicted all the variables and underpredicted the velocity due to the lack of sufficient mesh resolution. The computational time for the three meshes were 42.5 h (Mesh A), 23 h (Mesh B), and 17 h (Mesh C). Hence, the results indicate that the mesh size of 757,921 (Mesh B) provided a good compromise between computational cost and accuracy and was used for all subsequent simulations.

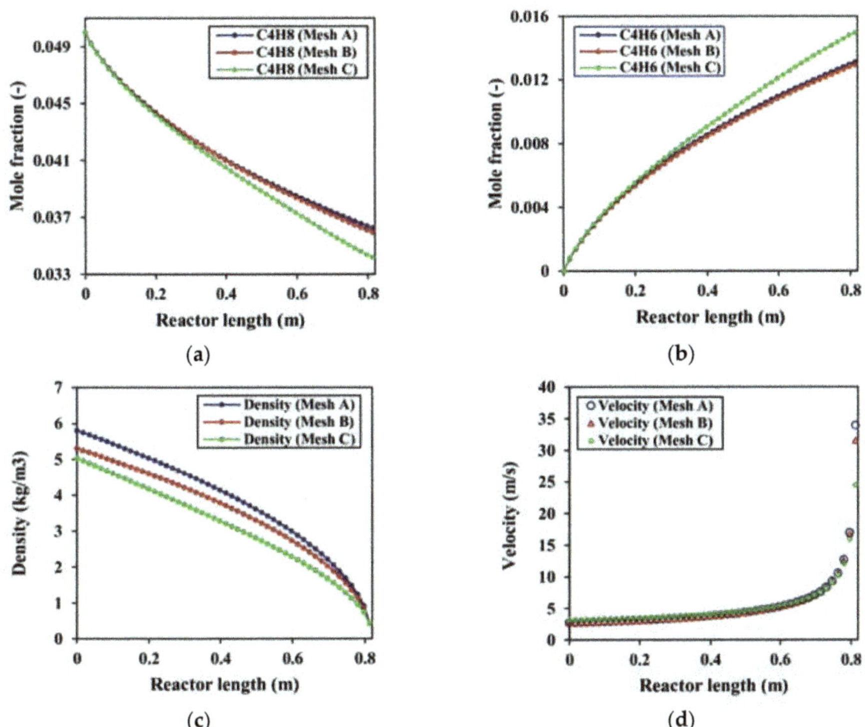

Figure 4. Mesh independency test results: (**a**) mole fraction of butene; (**b**) mole fraction of butadiene; (**c**) gas density; (**d**) gas velocity.

4.2. 3D Model Validation and Analysis

Figure 5 indicates the comparison between the CFD simulation results and experimental data in terms of mole fractions. Both the CFD simulation and experiment studies were carried out at 633 K under isothermal conditions achieved by suitable temperature control. Five different runs were made by varying the space time (residence time) from 0.069 h to 0.398 h. During the reactions, the butene and oxygen were consumed while producing butadiene, carbon dioxide, and water. To accurately validate the adequacy of the CFD model, the fundamental conservation equations and reaction kinetic equations derived based on the experimental results were employed. In addition, process conditions and model parameters used in the experiment (Table 3) were also used. It was assumed that the gravitational effect was negligible and the isothermal conditions were maintained within each tube as used in the experiment. In addition, the catalyst bed was modeled as a porous zone. As a result, it was observed that the data from the CFD simulation fitted well with the experimental results, giving a maximum error of 7.5%.

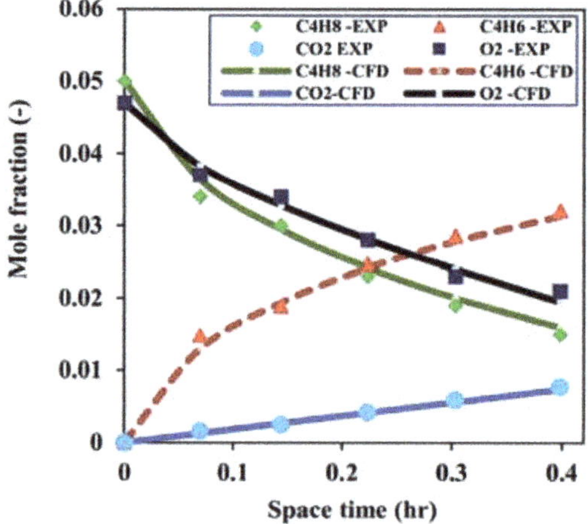

Figure 5. Comparison of the CFD simulation results with experimental data in terms of mole fractions (EXP: experiment).

In industrial processes, cylindrical fixed-bed reactors with relatively large diameter filled with solid catalysts are typically used for mild exothermic or endothermic reactions due to their simple design and straightforward construction. Meanwhile, if the reactions require extensive heat removal, tube-type reactors are appropriate for temperature control [36]. It is known that the ODH of butene to butadiene is a highly exothermic reaction where a substantial amount of heat should be removed during the reactions.

Therefore, Sterrett et al. [18] performed an adiabatic reactor operation with complete oxygen consumption to investigate the extent of exothermicity and the temperature rise in the reactor tube. Here, no temperature control of the reactor tubes was implemented. Figure 6 shows a comparative plot between the experimental adiabatic run and the CFD simulation. From the figure, the inlet temperature specified at 560 K gradually increased across the tube length until it reached a peak value of 763 K (experiment), and then gradually decreased to 727 K as a result of inevitable heat losses in the experimental setup. Consequently, a computer simulation was carried out by the same research group under the same adiabatic conditions from which an error of 2.53% was obtained. In comparison with the CFD adiabatic simulation, a higher error of 5.72% was realized. This is partly due to the lack of adequate information on the adiabatic experimental conditions such as flow rate and differences in thermophysical properties such as specific heat capacity of the catalyst and gases.

The results of this study also showed a similar upward trend with the experimental data, resulting in a final temperature of 720.06 K. The selectivity, as obtained from the experimental study, was 92%, whereas the computer simulation by Sterrett et al. [18] predicted the value as 96%. In this study, the selectivity was calculated to be 92.21% according to Equation (20), showing a strong agreement between the experimental results and CFD simulation.

$$\text{Selectivity} = 100 \times \frac{\text{moles of butadiene produced}}{\text{moles of butene reacted}} \tag{20}$$

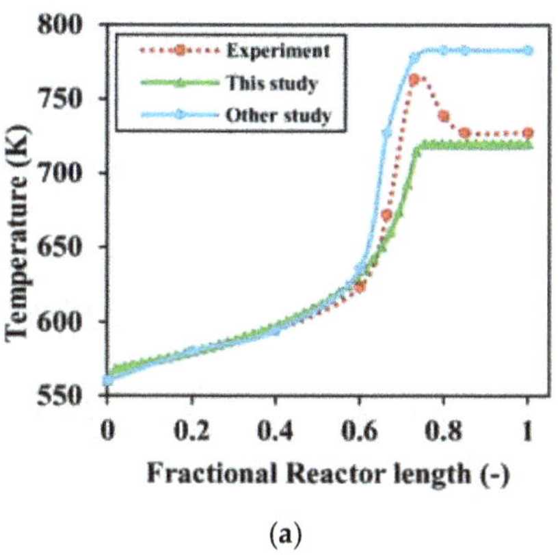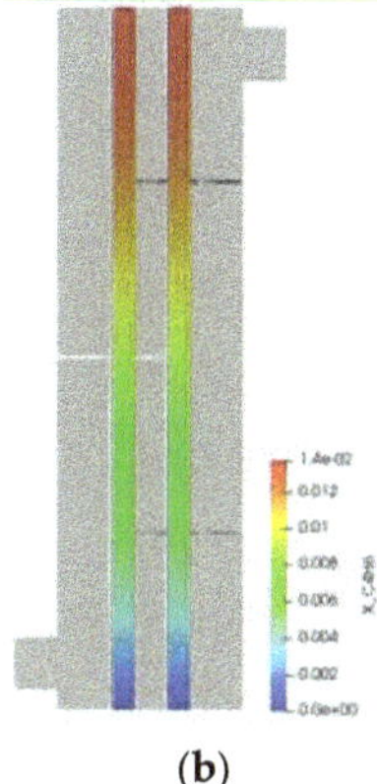

Figure 6. Validation of the CFD model under adiabatic conditions: (**a**) temperature comparison between CFD simulation, experimental data, and simulation reported by Sterrett et al. [18]; (**b**) concentration profiles of the CFD simulation.

4.3. Parametric Study

This section discusses the parametric study that involves the variation of operational parameters. Upon satisfactory validation of the CFD model, simulations under different thermal conditions such as isothermal, non-isothermal, and adiabatic were examined. Consequently, by varying the inlet temperature and velocity of the feed gas, the effects of these parameters on the reactor performance such as conversion (Equation (21)), selectivity (Equation (20)), and yield (Equation (22)) were evaluated.

$$\text{Conversion} = 100 \times \frac{\text{moles of butene reacted}}{\text{moles of butene fed}} \tag{21}$$

$$\text{Yield} = 100 \times \frac{\text{moles of butadiene produced}}{\text{moles of butadiene could be formed}} \tag{22}$$

4.3.1. Effect of Thermal Conditions

Based on the previous validation of the reactor model with the experimental results, concentration profiles of the reactants and products under different thermal conditions were predicted by the CFD model.

Figure 7 shows the results obtained by CFD simulation under isothermal conditions where the temperature of the reactor was maintained at 633 K. It can be seen that the mole fractions of reactants and products gradually decreased and increased along the reactor length. This is because the isothermal conditions inhibit the formation of hot spots in the reactor, thereby reducing the rate of reaction and consequent formation of side products. The butene conversion, yield, and selectivity of butadiene were estimated at values of 25.68%, 24.35% and 94.82%, respectively, of the two tubes that lay on the sectional plane of the longitudinal cross-sectional.

Conversely, another simulation was carried out under non-isothermal conditions where the shell-side coolant flowed around the tubes to remove the heat generated by the reactions. Water at 300 K was used as the shell-side coolant. Compared to the isothermal operation where a constant temperature was maintained inside the reactor tubes, the temperature in this case varied along the axial and radial directions of the tubes, resulting in a different concentration profile (Figure 8b). From Figure 8a, steeper concentration gradients were obtained in the non-isothermal simulation. Values of butene conversion and yield of butadiene were calculated to be 93.63% and 88.38%, respectively. Despite higher values of conversion and yield in this case, the selectivity of butadiene was estimated to be 94.40%, which was lower than that obtained in the isothermal simulation. This presupposes that higher temperatures would result in higher generation of side products; hence, the need for efficient temperature control within the reactor tubes. To reaffirm this assertion, an adiabatic simulation as well as further investigation of the reactor performance at various temperatures was executed.

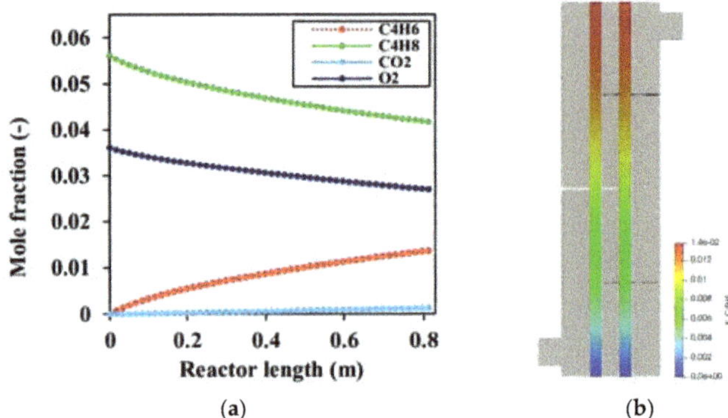

Figure 7. Mole fraction distribution and mole fraction contour plot under isothermal conditions: (**a**) mole fraction distributions of each species; (**b**) contour plot of butadiene mole fraction.

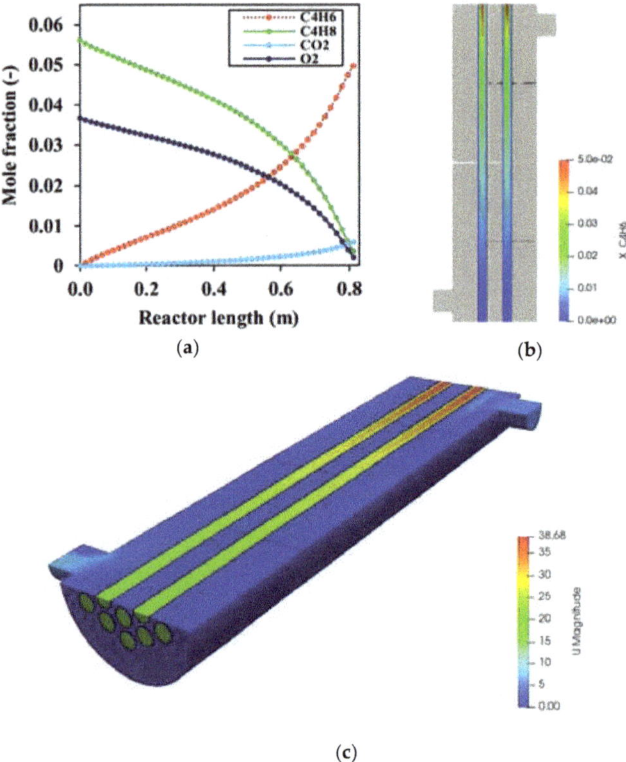

Figure 8. Mole fraction distribution and mole fraction contour plot under non-isothermal conditions: (**a**) mole fraction distributions of each species; (**b**) contour plot of butadiene mole fraction; (**c**) 3D velocity profile.

Figure 9 shows the results of the adiabatic operation in terms of species mole fractions. The simulation employed the operating conditions used in both the isothermal and non-isothermal studies above, in which the inlet temperature was fixed at 633 K and the inlet mole fractions of butene, oxygen, and water were 0.0563, 0.0367, and 0.717, respectively. During the reactions, the mole fractions of the reactants decreased considerably until the butene and oxygen were completely consumed, producing very high amounts of butadiene and carbon dioxide. After oxygen is depleted, the mole fractions of species remain constant as the oxygen acts as a limiting reactant. The conversion of butene and yield of butadiene in this case were calculated to be 98.41% and 92.81%, respectively. The selectivity of butadiene was also computed as 94.31%.

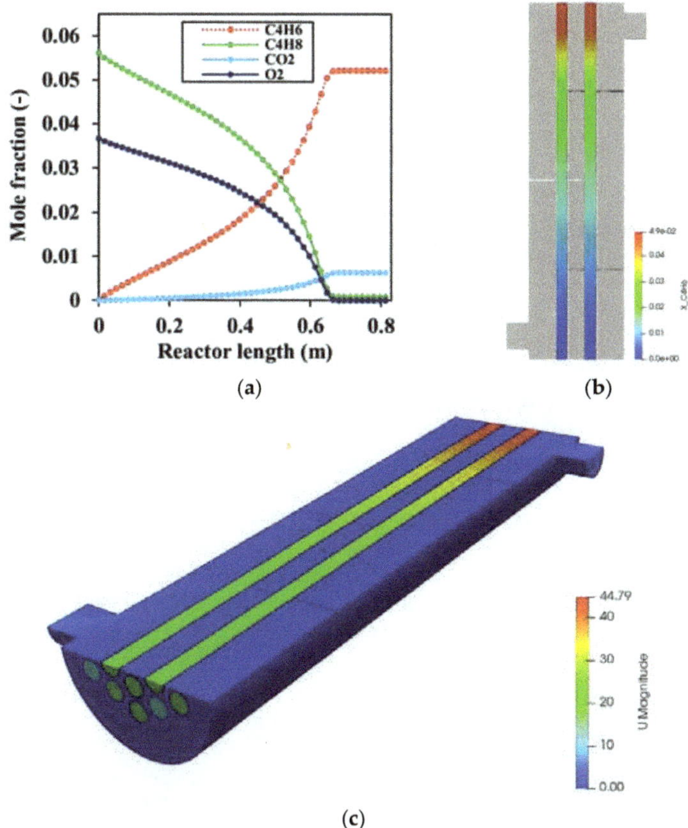

Figure 9. Mole fractions distribution and mole fraction contour plot under adiabatic conditions: (a) mole fraction distributions of each species; (b) contour plot of butadiene mole fraction; (c) 3D velocity profile.

Figure 10a,b shows the temperature contour of the reactor tubes under adiabatic and non-isothermal conditions with their accompanying Cartesian plot shown in Figure 10c. It was observed that under the adiabatic conditions (no cooling), the reaction temperature increased to 862 K until all the limiting reactant was completely depleted. This confirms the extent of exothermicity of the ODH reaction and calls for effective and efficacious temperature control. As shown in Figure 10a,b, a temperature hotspot was formed close to the reactor outlet and hence provides a new insight into selecting the best methods (counter or co-current heat transfer) for controlling the reaction temperature. Similarly, Figure 10c shows a gradual rise in temperature across the reactor tube length (0 to 0.6 m) and rises sharply afterward (0.6 to 0.8128 m), reiterating the need to investigate the best methods for controlling the reaction temperature (not shown in this study).

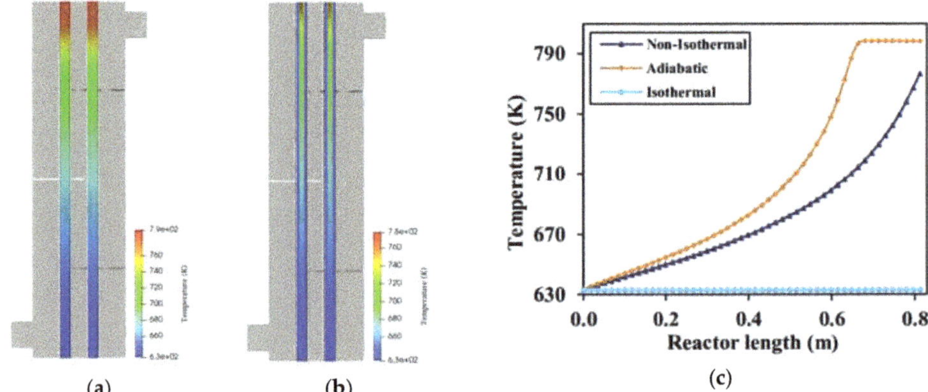

Figure 10. Temperature contour and profiles of a multi-tubular reactor under different thermal conditions: (**a**) temperature contour under adiabatic conditions; (**b**) temperature contour non-isothermal conditions (**c**) temperature profiles under isothermal, non-isothermal, and adiabatic conditions.

4.3.2. Effect of Temperature

Despite the different studies on reactor performance under the various thermal conditions (isothermal, non-isothermal, and adiabatic), it is important to investigate the effect of varying temperatures on the generation and consumption of the species to provide sufficient insight into modeling, optimization, and scale-up of multi-tubular reactors. Although the assumption of isothermal conditions as used in this parametric study deviated slightly from industrial application, it provides to a large extent, adequate information on the reactor performance for possible adoption and implementation.

To this end, the effect of gas temperature on the reactor performance was carried out by varying the reactor temperature from 610 to 690 K at a constant gas velocity of 35.31 m/s. Figure 11 shows the distribution of mole fractions for the various species along the reactor length. From the figure, butene and oxygen were gradually consumed with decreasing slopes. At the product side, butadiene and carbon dioxide were also generated with increasing slopes. It can be noted that carbon dioxide production increases linearly with temperature. This means that the relative amount of carbon dioxide produced compared to butadiene increases as the temperature increases. As expected, it can be seen that the extent of the reaction increases as the temperature rises. This can be explained partly by the kinetic theory of gases where higher temperatures give rise to more rapid collisions between molecules for faster reactions to take place and partly by the temperature dependence of the Arrhenius equation.

However, it is worth knowing that side reactions that result in lower selectivity are likely to occur when the temperature increases. Figure 12 shows the reactor outlet concentrations and reactor performance indices (conversion, yield and selectivity) as functions of varying temperatures. The conversion of butene and yield of butadiene show an upward increase in value (positive gradients) with maximum values of 86.02% and 81.26% at 690 K. From Figure 12b, it can be seen that the yield and conversion gradients increased gradually from 610 to 633 K, slightly sharply from 640 K to 670 K, and very sharply from 670 K to 690 K. However, in a separate analysis (results not shown), it was observed that the gradient started to decrease at 700 K, although the conversion and yield face values increased. This revelation provides an insight into the possible optimization of the reactor at these temperatures (670 K to 690 K). The selectivity of butadiene, however, slightly increased from 610 K to 633 K and then steeply decreased afterward. This can be explained by the sudden change in carbon dioxide concentration gradient in Figure 12a, where the gradient remained fairly constant from 610 K to 633 K and increased sharply thereafter. This shows that the selectivity is inversely proportional to the yield and conversion. As such, a compromise needs to be made in selecting the appropriate values of these performance indices.

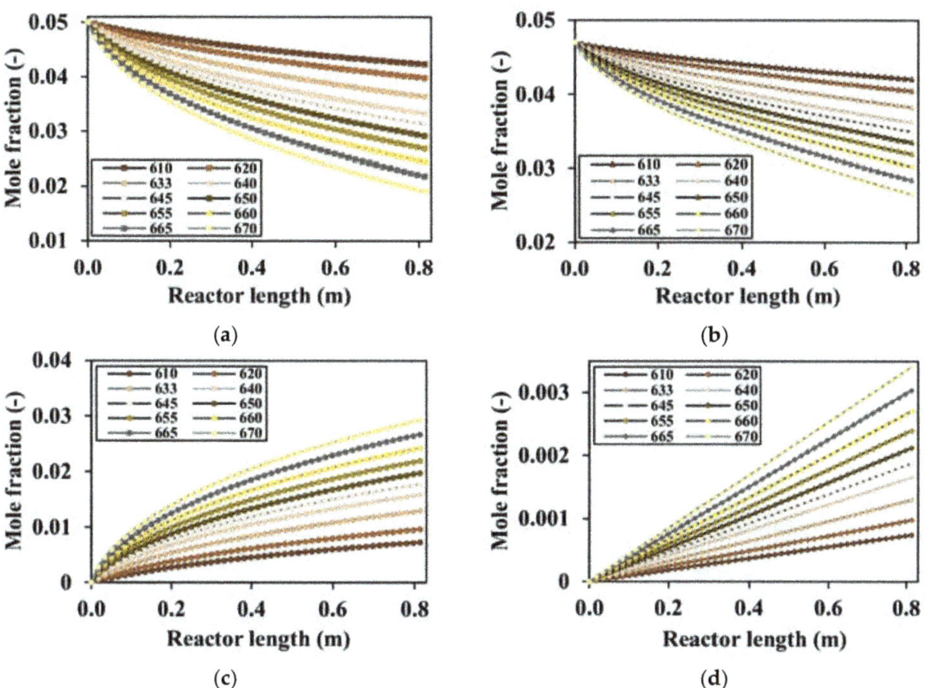

Figure 11. Mole fraction distribution along the reactor length at the temperature range between 610 K and 670 K: (**a**) butene; (**b**) oxygen; (**c**) butadiene; (**d**) carbon dioxide.

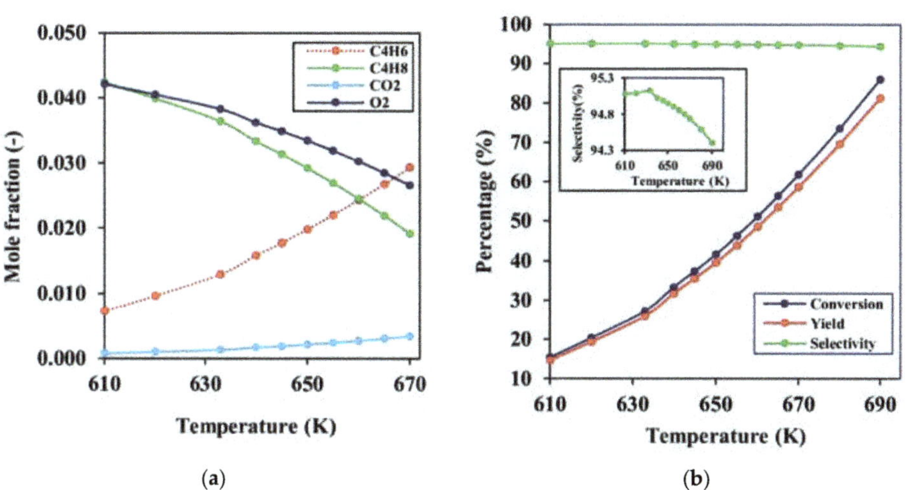

Figure 12. Mole fractions and reactor performance profiles as a function of temperature: (**a**) mole fractions; (**b**) conversion, yield, and selectivity.

4.3.3. Effect of Gas Velocity

Figure 13 illustrates the relationship between the mole fractions and gas velocities from 3.25 to 35.31 m/s. High reaction rates were obtained at low velocities and low reaction rates at high velocities. This phenomenon can best be elucidated with the concept of residence time, where, as the gas velocity increases, the residence time (space time) decreases, thereby reducing the time available for the interaction between the catalyst particles and reactant species. Additionally, as shown in Figure 10c, the rate of reaction increased gradually across the reactor and therefore requires a longer residence time for efficient conversion of reactants to products. This is further demonstrated in Figure 14a where the concentration gradients remained nearly linear from 35.31 to 10.92 m/s and changed exponentially from 8.04 to 3.25 m/s. In the same Figure 14b, the reactor performance indices (conversion, yield, and selectivity) were plotted against the gas velocities. It can be clearly seen that as the gas velocity increases, selectivity also increases whilst the converse was observed for the yield and conversion. This observation is due to the fact that an increase in residence time (low velocity) increases the reaction rate and consequently, the conversion and yield. From Figure 13d, the production of carbon dioxide (a side product) was observed to increase linearly with decreasing gas velocities. Hence, as the velocity of the gas increases (reduction in residence time), the carbon dioxide production decreases, resulting in an increase in selectivity (relatively high production of butadiene). Therefore, a good optimization of these reactor performance indices is required to adequately choose the best performing process conditions.

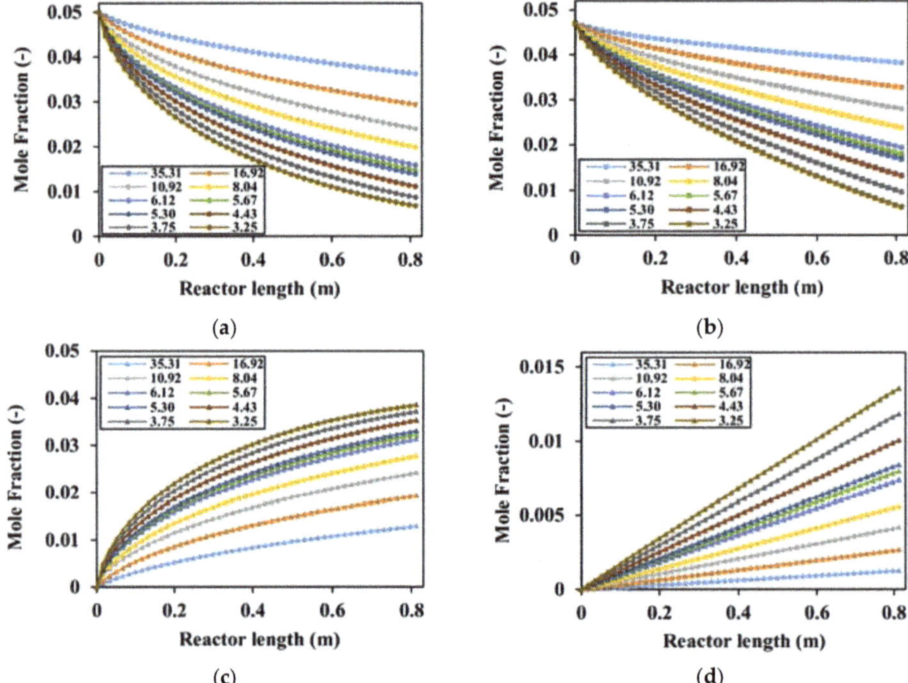

Figure 13. Mole fraction distribution along the reactor length at the temperature range between 3.25 m/s and 35.31 m/s: (**a**) butene; (**b**) oxygen; (**c**) butadiene; (**d**) carbon dioxide.

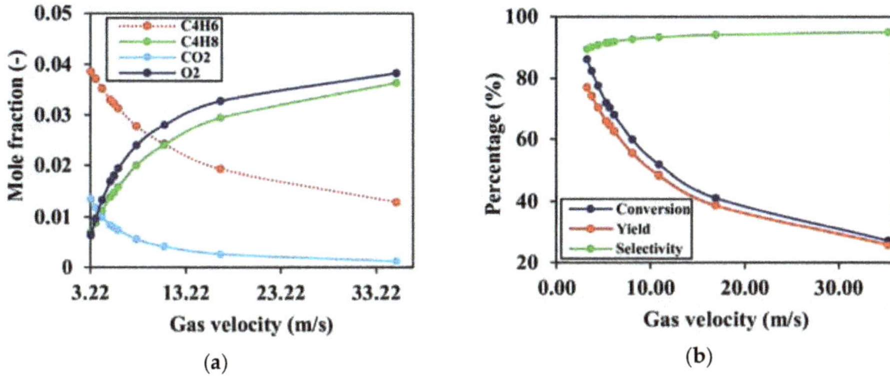

Figure 14. Mole fractions and reactor performance profiles as a function of gas velocity: (**a**) mole fractions; (**b**) conversion, yield, and selectivity.

5. Conclusions

In this study, the oxidative dehydrogenation of butene to 1,3-butadiene over a Zn–Cr–Fe catalyst in a 3-dimensional multi-tubular shell and tube type reactor was investigated. Rigorous mathematical modeling of reaction kinetics from the experimental work of Sterrett et al. [17] was implemented. The catalyst was modeled as a porous zone to include the effects of pressure drop, mass, and heat transfer. The shell-side flow was modeled with the standard k–ε turbulent model whilst the tube side was assumed to be laminar. To ensure effective cooling of the reactor tubes, water at 300 K was injected as the shell-side coolant in a counter-current fashion.

Results from the developed CFD model agreed well with the experimental data, giving a maximum error of 7.5%. Three parametric studies were carried out to investigate the effect of thermodynamic conditions, temperature, and flow rate on key reactor performance indices such as conversion, yield, and selectivity. In the first parametric study, a CFD simulation was executed for three thermodynamic reactor conditions (isothermal, non-isothermal, and adiabatic). The isothermal condition gave the least conversion and yield (25.68% and 24.35%), but also the highest selectivity of 94.82%. Under the non-isothermal condition, a hot spot was formed at the top most part of the reactor tubes (0.7 to 0.8128 m) with considerable increase in yield and conversion (93.63% and 88.38%), but a decrease in selectivity of 94.40%. Similar to the non-isothermal state, the adiabatic temperature was found to rise from 633 K to 860 K until all the limiting reactant (O_2) was completely consumed. The selectivity further decreased in this case while both yield and conversion increased to a maximum (98.41% and 92.81%).

It was observed and can be concluded from this study that the temperature rises slowly across the reactor (0 to 0.6 m), but increases sharply, thereafter forming a hot spot. Hence, to efficiently control the reactor temperature, careful design of the shell-side for maximum heat removal close to the outlet should be considered. It is recommended that, in lieu of the cooling water, a more efficient coolant should be used for effective temperature control.

In the second and third parametric studies, the effect of varying flow rates and temperature under isothermal conditions were investigated. From the results, it was clearly seen that temperature significantly affects the rate of reactions as well as the reactor performance. As the temperature increased, the conversion and yield also increased (from 15.35% to 86.02% and 14.61% to 81.26%), but with a consequential reduction in selectivity (from 95.08% to 94.13%). Additionally, an increase in velocity favored the relative production of butadiene with a maximum selectivity of 95.12% obtained at 35.31 m/s. The generation of side product (CO_2) was, however, favored at low velocities due to the increase in residence time since the CO_2 kinetics is zero-order with respect to butene and butadiene. Hence, as the velocity decreased, the rate of CO_2 production increased, thereby decreasing the selectivity of butadiene production.

Author Contributions: Conceptualization, J.M., D.Q.G. and S.H.; Methodology, J.M. and D.Q.G. Software, J.M. and D.Q.G.; Validation, J.M., D.Q.G., and S.H.; Formal analysis, J.M., D.Q.G. and S.H.; Investigation, J.M. and D.Q.G.; Resources, J.M., D.Q.G., and S.H.; Data curation, J.M. and D.Q.G.; Writing—original draft preparation, J.M. and D.Q.G.; Writing—review and editing, J.M., D.Q.G., and S.H.; visualization, J.M. and D.Q.G.; Supervision, S.H.; Project administration, S.H.; Funding acquisition, S.H. All authors have read and agreed to the published version of the manuscript.

Funding: This research was funded by a National Research Foundation of Korea (NRF) grant funded by the Korea government (MSIT) (No. 62163-01). This research was also supported by the Korea Institute for Advancement of Technology (KIAT) grant funded by the Korea Government (MOTIE, P0008475, Development Program for Smart Digital Engineering Specialist).

Conflicts of Interest: The authors declare no conflict of interest. The funders had no role in the design of the study; in the collection, analyses, or interpretation of data; in the writing of the manuscript, or in the decision to publish the results.

References

1. Rischard, J.; Antinori, C.; Maier, L.; Deutschmann, O. Oxidative dehydrogenation of n-butane to butadiene with Mo-V-MgO catalysts in a two-zone fluidized bed reactor. *Appl. Catal. A Gen.* **2016**, *511*, 23–30. [CrossRef]
2. White, W.C. Butadiene production process overview. *Chem. Biol. Interact.* **2007**, *166*, 10–14. [CrossRef] [PubMed]
3. Qi, Y.; Liu, Z.; Liu, S.; Cui, L.; Dai, Q.; He, J.; Dong, W.; Bai, C. Synthesis of 1,3-butadiene and its 2-substituted monomers for synthetic rubbers. *Catalysts* **2019**, *9*, 97. [CrossRef]
4. Yoshimura, Y.; Kijima, N.; Hayakawa, T.; Murata, K.; Suzuki, K.; Mizukami, F.; Matano, K.; Konishi, T.; Oikawa, T.; Saito, M.; et al. Catalytic cracking of naphtha to light olefins. *Catal. Surv. Japan* **2001**, *4*, 157–167. [CrossRef]
5. Yan, W.; Kouk, Q.Y.; Luo, J.; Liu, Y.; Borgna, A. Catalytic oxidative dehydrogenation of 1-butene to 1,3-butadiene using CO_2. *Catal. Commun.* **2014**, *46*, 208–212. [CrossRef]
6. Ren, T.; Patel, M.; Blok, K. Olefins from conventional and heavy feedstocks: Energy use in steam cracking and alternative processes. *Energy* **2006**, *31*, 425–451. [CrossRef]
7. Hong, E.; Park, J.H.; Shin, C.H. Oxidative Dehydrogenation of n-Butenes to 1,3-Butadiene over Bismuth Molybdate and Ferrite Catalysts: A Review. *Catal. Surv. Asia* **2016**, *20*, 23–33. [CrossRef]
8. Bhasin, M.M.; McCain, J.H.; Vora, B.V.; Imai, T.; Pujadó, P.R. Dehydrogenation and oxydehydrogenation of paraffins to olefins. *Appl. Catal. A Gen.* **2001**, *221*, 397–419. [CrossRef]
9. Herguido, J.; Menendez, M.; Santamaria, J. Oxidative Dehydrogenation of n -Butane in a Two-Zone Fluidized-Bed Reactor. *Ind. Eng. Chem. Res.* **1999**, *38*, 90–97. [CrossRef]
10. Schäfer, R.; Noack, M.; Kölsch, P.; Stöhr, M.; Caro, J. Comparison of different catalysts in the membrane-supported dehydrogenation of propane. *Catal. Today* **2003**, *82*, 15–23. [CrossRef]
11. Park, S.; Lee, Y.; Kim, G.; Hwang, S. Production of butene and butadiene by oxidative dehydrogenation of butane over carbon nanomaterial catalysts. *Korean J. Chem. Eng.* **2016**, *33*, 3417–3424. [CrossRef]
12. Callejas, C.; Soler, J.; Herguido, J.; Menéndez, M.; Santamaría, J. Catalytic dehydrogenation of n-butane in a fluidized bed reactor with separate coking and regeneration zones. *Stud. Surf. Sci. Catal.* **2000**, *130*, 2717–2722. [CrossRef]
13. Vajda, S.; Pellin, M.J.; Greeley, J.P.; Marshall, C.L.; Curtiss, L.A.; Ballentine, G.A.; Elam, J.W.; Catillon-Mucherie, S.; Redfern, P.C.; Mehmood, F.; et al. Subnanometre platinum clusters as highly active and selective catalysts for the oxidative dehydrogenation of propane. *Nat. Mater.* **2009**, *8*, 213–216. [CrossRef]
14. Rischard, J.; Franz, R.; Antinori, C.; Deutschmann, O. Oxidative dehydrogenation of butenes over Bi-Mo and Mo-V based catalysts in a two-zone fluidized bed reactor. *AIChE J.* **2017**, *63*, 43–50. [CrossRef]
15. Elkhalifa, E.A.; Friedrich, H.B. Magnesium oxide as a catalyst for the dehydrogenation of n-octane. *Arab. J. Chem.* **2018**, *11*, 1154–1159. [CrossRef]
16. Xingan, W.; Huiqin, L. Comparison of the technology of oxidative dehydrogenation in a fluidized-bed reactor with those of other reactors for butadiene. *Ind. Eng. Chem. Res.* **1996**, *35*, 2570–2575. [CrossRef]

17. Hiwale, R.; Smith, R.; Hwang, S. A novel methodology for the modeling of CO_2 absorption in monoethanolamine (MEA) using discrimination of rival kinetics. *J. Ind. Eng. Chem.* **2015**, *25*, 78–88. [CrossRef]

18. Sterrett, J.S.; Mcllvried, H.G. Kinetics of the Oxidative Dehydrogenation of Butene to Butadiene over a Ferrite Catalyst. *Ind. Eng. Chem. Process. Des. Dev.* **1974**, *13*, 54–59. [CrossRef]

19. Ding, X.J.; Xiao, D.L.; Wang, X.L.; Liu, S.L. The redox model of the kinetics for the oxidative dehydrogenation over the ferrite catalyst. *J. Mol. Catal.* **1988**, *2*, 25–30.

20. Téllez, C.; Menéndez, M.; Santamaría, J. Kinetic study of the oxidative dehydrogenation of butane on V/MgO catalysts. *J. Catal.* **1999**, *183*, 210–221. [CrossRef]

21. Ajayi, B.P.; Abussaud, B.; Jermy, R.; Al Khattaf, S. Kinetic modelling of n-butane dehydrogenation over CrOxVOx/MCM-41 catalyst in a fixed bed reactor. *Prog. React. Kinet. Mech.* **2014**, *39*, 341–353. [CrossRef]

22. Madaan, N.; Haufe, R.; Shiju, N.R.; Rothenberg, G. Oxidative Dehydrogenation of n-Butane: Activity and Kinetics over VO_x/Al_2O_3 Catalysts. *Top. Catal.* **2014**, *57*, 1400–1406. [CrossRef]

23. Tanimu, A.; Jaenicke, S.; Alhooshani, K. Heterogeneous catalysis in continuous flow microreactors: A review of methods and applications. *Chem. Eng. J.* **2017**, *327*, 792–821. [CrossRef]

24. Tanimu, G.; Abussaud, B.A.; Asaoka, S.; Alasiri, H. Kinetic Study on n-Butane Oxidative Dehydrogenation over the (Ni, Fe, Co)-Bi-O/Î-Al_2O_3 Catalyst. *Ind. Eng. Chem. Res.* **2020**, *59*, 2773–2780. [CrossRef]

25. Mendoza, J.A.; Hwang, S. Tubular reactor design for the oxidative dehydrogenation of butene using computational fluid dynamics (CFD) modeling. *Korean J. Chem. Eng.* **2018**, *35*, 2157–2163. [CrossRef]

26. De Smet, C.R.H.; De Croon, M.H.J.M.; Berger, R.J.; Marin, G.B.; Schouten, J.C. Design of adiabatic fixed-bed reactors for the partial oxidation of methane to synthesis gas. Application to production of methanol and hydrogen-for-fuel-cells. *Chem. Eng. Sci.* **2001**, *56*, 4849–4861. [CrossRef]

27. Soler, J.; Téllez, C.; Herguido, J.; Menéndez, M.; Santamaría, J. Modelling of a two-zone fluidised bed reactor for the oxidative dehydrogenation of n-butane. *Powder Technol.* **2001**, *120*, 88–96. [CrossRef]

28. Hukkanen, E.J.; Rangitsch, M.J.; Witt, P.M. Non-adiabatic multitubular fixed-bed catalytic reactor model coupled with shell-side coolant CFD model. *Ind. Eng. Chem. Res.* **2013**, *52*, 15437–15446. [CrossRef]

29. Huang, K.; Lin, S.; Wang, J.; Luo, Z. Numerical evaluation on the intraparticle transfer in butylene oxidative dehydrogenation fixed-bed reactor over ferrite catalysts. *J. Ind. Eng. Chem.* **2015**, *29*, 172–184. [CrossRef]

30. Dixon, A.G.; Nijemeisland, M. CFD as a design tool for fixed-bed reactors. *Ind. Eng. Chem. Res.* **2001**, *40*, 5246–5254. [CrossRef]

31. Chen, X.; Dai, J.; Luo, Z. CFD modeling using heterogeneous reaction kinetics for catalytic dehydrogenation syngas reactions in a fixed-bed reactor. *Particuology* **2013**, *11*, 703–714. [CrossRef]

32. Fattahi, M.; Kazemeini, M.; Khorasheh, F.; Darvishi, A.; Rashidi, A.M. Fixed-bed multi-tubular reactors for oxidative dehydrogenation in ethylene process. *Chem. Eng. Technol.* **2013**, *36*, 1691–1700. [CrossRef]

33. Launder, B.E.; Spalding, D.B. The numerical computation of turbulent flows. *Comput. Methods Appl. Mech. Eng.* **1974**, *3*, 269–289. [CrossRef]

34. Byrne, R.C. *TEMA Standards of the Tubular Exchanger*, 9th ed.; Tubular Exchanger Manufacturers Association, INC.: New York, NY, USA, 2007; pp. 298–300.

35. Gbadago, D.Q.; Oh, H.T.; Oh, D.H.; Lee, C.H.; Oh, M. CFD simulation of a packed bed industrial absorber with interbed liquid distributors. *Int. J. Greenh. Gas. Control.* **2020**. [CrossRef]

36. Worstell, J. *Adiabatic Fixed-Bed Reactors: Practical Guides in Chemical Engineering*; Butterworth-Heinemann: Texas, TX, USA, 2014.

Publisher's Note: MDPI stays neutral with regard to jurisdictional claims in published maps and institutional affiliations.

 *chemengineering* 

Article

An Assessment of Drag Models in Eulerian–Eulerian CFD Simulation of Gas–Solid Flow Hydrodynamics in Circulating Fluidized Bed Riser

Mukesh Upadhyay, Ayeon Kim, Heehyang Kim, Dongjun Lim and Hankwon Lim *

School of Energy and Chemical Engineering; Ulsan National Institute of Science and Technology (UNIST); 50 UNIST-gil, Eonyang-eup, Ulju-gun, Ulsan 44919, Korea; mukesh0223@unist.ac.kr (M.U.); okay1130@unist.ac.kr (A.K.); khrong@unist.ac.kr (H.K.); dongjun1993@unist.ac.kr (D.L.)
* Correspondence: hklim@unist.ac.kr; Tel.: +82-52-217-2935

Received: 29 April 2020; Accepted: 4 June 2020; Published: 6 June 2020

Abstract: Accurate prediction of the hydrodynamic profile is important for circulating fluidized bed (CFB) reactor design and scale-up. Multiphase computational fluid dynamics (CFD) simulation with interphase momentum exchange is key to accurately predict the gas-solid profile along the height of the riser. The present work deals with the assessment of six different drag model capability to accurately predict the riser section axial solid holdup distribution in bench scale circulating fluidized bed. The difference between six drag model predictions were validated against the experiment data. Two-dimensional geometry, transient solver and Eulerian–Eulerian multiphase models were used. Six drag model simulation predictions were discussed with respect to axial and radial profile. The comparison between CFD simulation and experimental data shows that the Syamlal-O'Brien, Gidaspow, Wen-Yu and Huilin-Gidaspow drag models were successfully able to predict the riser upper section solid holdup distribution with better accuracy, however unable to predict the solid holdup transition region. On the other hand, the Gibilaro model and Helland drag model were successfully able to predict the bottom dense region, but the upper section solid holdup distribution was overpredicted. The CFD simulation comparison of different drag model has clearly shown the limitation of the drag model to accurately predict overall axial heterogeneity with accuracy.

Keywords: circulating fluidized bed riser; computational fluid dynamics; eulerian–eulerian; drag models; 2D simulation

1. Introduction

The gas–solid fluidized bed reactors are widely used in the petrochemical, power generation, environmental and metallurgical industries [1]. In the circulating fluidized bed (CFB) system, solid particles are separated from the fluid stream using cyclone and recycle to the bed. The solid particles fluid-like state provide excellent heat and mass transfer characteristic and circulation of solid offer operational flexibility [2,3]. The CFB riser section serves as the main reaction zone, in which strands of particles (clusters) influence the flow and the performance of reactor [4]. Over the years, considerable effort has been made to better design and operation of CFB reactor. In particular, focus to better understand the underlying gas–solid flow hydrodynamics in the riser section.

Over the year, computational fluid dynamics (CFD) domain has undergone significant development and regularly applied for investigating gas-solid multiphase flow phenomena. For example, direct numerical simulation (DNS), large eddy simulation (LES) and Reynolds-Averaged Navier-Stokes simulation (RANS) are three modelling methods at different computational scales to account the turbulence in multiphase flow. The accuracy of these numerical methods depends on the scheme. Several high order schemes were developed, such as essentially non-oscillatory (ENO) method [5,6], weighted ENO (WENO) method [7], discontinuous Galerkin (DG) method [8], radial basis function method [9] and gas-kinetic method [10,11]. However, for most industrial scale problem DNS and LES simulation were considered impractical for the general-purpose design tool [12]. RANS based CFD method has become a standard platform to simulate gas–solid multiphase flow in industrial scale fluidized bed reactor [13]. To model the gas–solid flow, the most commonly employed CFD simulation methods are the Eulerian–Eulerian two-fluid model (EE-TFM), [14–16] Eulerian–Lagrangian discrete element method (EL-DEM) [17,18], and hybrid multiphase particle-in-cell (MP-PIC) [19,20]. The principal difference among these approaches is on the treatment of the secondary solid phase. In EE-TFM approach kinetic theory of granular flow (KTGF) closure model were incorporated to simulate solid particle flow. EE-TFM model constitutive relationship contains solid phase stress, interphase momentum transfer and particle interactions [13,16,21]. Researchers have shown that the interphase drag force is the critical closure model to accurately predict the gas–solid flows [22,23]. Agrawal et al. [23] has shown that the solid phase stress contribution was insignificant, and it is crucial to accurately account for the solid particle clusters.

Researchers have proposed the drag model based on flow conditions. In the literature, drag models were classically arranged into two groups, i.e., structure-based drag models and conventional drag models. Among the conventional drag models, Wen-Yu [24], Gidaspow [25] and Syamlal-O'Brien [26] were widely incorporated, which were derived from dense packed bed experimental pressure drop data and terminal velocity deduce using single solid particle. The most widely used Gidaspow model correlation consist of Ergun [27] and Wen-Yu [24] equations. To accurately simulate the riser gas-solid multiphase flow, researchers incorporated drag coefficient to account for the particle cluster. Researchers have pointed out that the particles cluster settling velocity is about 20% to 100% higher than the single particles. Also recognized that the such drag reduction occurred in the low solid volume fraction region. Various drag models have been incorporated to simulate the gas–solid riser flow, as detail were provided in Appendix A. Among the works in the literature, Gidaspow and EMMS drag model has been widely used. Shah et al. [28] simulate fluid catalytic cracking (FCC) riser section using EMMS and the Gidaspow drag model for low and high solid circulation rate. They reported EMMS drag model successfully able to predict the riser axial heterogenous profile. However, unable to predict the solid volume fraction transition region. Vaishali et al. [29] perform CFD simulation for fast fluidization and dilute phase transport flow regime with Geldart B particles. They reported that the Wen-Yu drag model underpredicts the velocity profile and Syamlal-O'Brien drag model show better prediction. Wang et al. [30] investigate the models and model parameters variation on hydrodynamic prediction behavior for high solid circulation rate. Their drag model comparison study shows that the Syamlal-O'Brien, Gidaspow and Wen-Yu drag models predict the global profile similarly; however, on radial direction, the Syamlal-O'Brien drag model shows better accuracy. Similarly, Zhang et al. [31,32] compare EMMS and Gidaspow drag models, and it was reported that the EMMS drag model prediction is good in the axial and radial direction. Despite several studies, it is important to examine the influence of drag models for specific flow condition and material type. Hence, in the present work an assessment of the appropriate drag model selection for in-house CFB riser setup was carried out. Generated CFD simulation profile were plotted against experimentally obtain axial solid holdup data for detail assessment and validation purpose.

In the present work six drag models (Syamlal-O'Brien [26], Gidaspow [25], Wen-Yu [24], Huilin-Gidaspow [33], Gibilaro [34] and Helland [35]) were investigated. In order to keep the computational load at a reasonable level, simple 2D simulations were performed. A detailed discussion is presented to illustrate the inherent difference between different drag model prediction. Such a comparison study is expected to provide clear guidelines on selection of appropriate drag model selection for gas–solid riser flow.

2. Simulation System

In the present work, air used as fluidizing gas and 4 kg of silica sand solid particles were loaded into CFB reactor. CFD simulation was conducted at the same flow conditions as the work described in previous paper [16]. CFB riser simplified two-dimensional computational domain with the height of 3 m and a diameter of 0.025 m, shown in Figure 1a. To ensure symmetry flow the gas velocity was specified at the bottom, the solid phase were introduced from two sides and straight outlet, shown in Figure 1b. The gas and solid particle properties, system geometry and flow condition were summarized in Table 1. Our previous study investigated the optimum number of computational cells [16]. Therefore, 19,500 (13 × 1500) cells with a size of a 15-particle diameter were used to simulate the gas–solid flow in the riser.

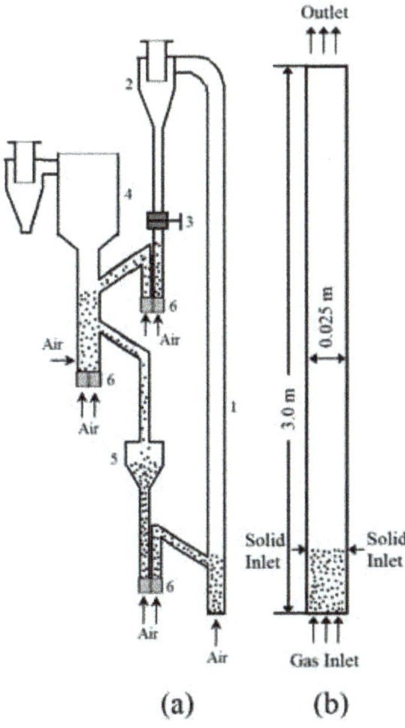

Figure 1. (**a**) Schematic diagram (**a**) CFB reactor (**b**) simplified 2D computational domain of CFB riser (1 —riser bed; 2—cyclone; 3—ball valves; 4—bubbling bed; 5—recirculation region; 6—loop-seal).

Table 1. System geometry, gas and solid phase properties and flow condition.

Description	Value
CFB riser:	
Diameter, D	0.025 m
Height, H	3.0 m
Fluidizing media properties:	
Gas density, ρ_g	1.225 kg/m^3
Solid phase density, ρ_s	2525 kg/m^3
Particle mean diameter, d_p	130 μm
Flow condition:	
Superficial gas velocity, U_g	2 m/s
Solid particle flux, G_s	39.14 kg/m^{2}s

3. CFD Model

Gas–solid CFD simulation was developed based on Eulerian–Eulerian two-fluid model (EE-TFM) [16] in ANSYS Fluent 2020R1 software. EE-TFM simulation equations are presented in Appendix B, and a detailed description of governing equations can be found elsewhere [14,15]. To correctly capture the riser gas–solid multiphase flow phenomena, granular temperature equation in partial-differential form was used to resemble the solid phase properties [36]. The standard k-ε turbulence model was selected to account the gas phase turbulence. Under Johnson and Jackson [16], the particle–wall interaction parameter, specularity coefficient (φ) and particle–wall restitution coefficient (e_w) value were set at 0.0001 and 0.9, respectively. Elasticity between solid phase, defined as the particle–particle restitution coefficient (e_{ss}), was set to 0.9. The transient CFD calculations were performed, where time step of 0.0005 s was selected to capture riser gas-solid flow behavior. Coupling between the velocity and pressure was employed by a phase-coupled SIMPLE scheme, and other important simulation parameters are given in Table 2. Before post-processing, the riser outlet mass flow rate was monitored as a function of time to confirm the pseudo-steady state profile, as shown in Figure 2. All the transient simulations were run for 30 s and reported results were employed by time-averaging the simulation results over last 20 s. The interphase momentum exchange between the gas and solid phases was provided through the drag coefficient. Six different drag models investigated in the present work; drag correlations are given in Table 3.

Table 2. Modeling parameters.

Particle–Wall and Particle–Particle Interactions Parameter Value:		
Specularity coefficient (φ)	-	0.0001
Particle–wall restitution coefficient (e_w)	-	0.9
Particle–particle restitution coefficient (e_{ss})	-	0.9
Packing limit ($\alpha_{s,max}$)	-	0.63
Transient solver calculation and convergence criteria:		
Time step (s)	-	0.0005
Convergence criteria	-	10^{-3}
Maximum iterations per time step	-	50
Discretization schemes settings:		
Momentum	-	1st order upwind
Volume fraction	-	1st order upwind
Transient formulation	-	1st order implicit

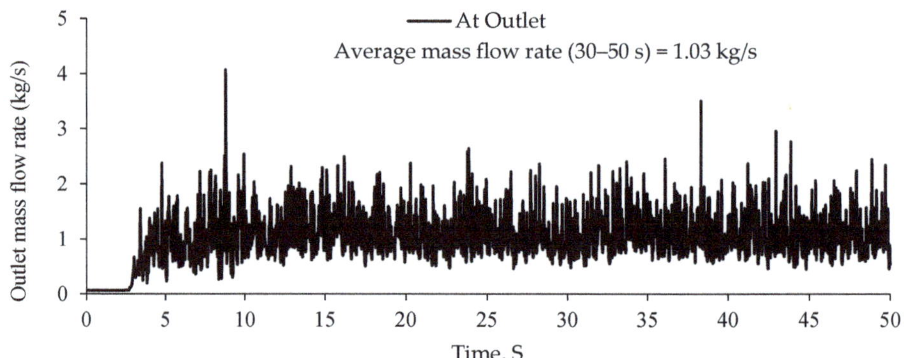

Figure 2. CFB riser outlet mass flow rate as a function of time.

Table 3. Drag model correlations.

---

1. Syamlal-O'Brien: [26]

$$K_{sg} = \frac{3\alpha_s\alpha_g\rho_g}{4V_{r,s}^2 d_s} C_D\left(\frac{Re_s}{V_{r,s}}\right)\left|\vec{V}_s - \vec{V}_g\right|$$

where

$$C_D = \left(0.63 + \frac{4.8}{\sqrt{\frac{Re_s}{V_{r,s}}}}\right)^2$$

$$V_{r,s} = 0.5\left(A - 0.06Re_s + \sqrt{(0.06Re_s)^2 + 0.12Re_s(2B - A) + A^2}\right)$$

$$A = \alpha_g^{4.14}$$

$$B = 0.8\alpha_g^{1.28} \; for \; \alpha_g \le 0.85$$

$$B = \alpha_g^{2.65} \;\; for \; \alpha_g > 0.85$$

$$Re_s = \frac{\rho_g d_s \left|\vec{V}_s - \vec{V}_g\right|}{\mu_g}$$

---

2. Wen-Yu: [24]

$$K_{sg} = \tfrac{3}{4}C_D \frac{\alpha_s\alpha_g\rho_g\left|\vec{V}_s - \vec{V}_g\right|}{d_s}\alpha_g^{-2.65}$$

where

$$C_D = \frac{24}{\alpha_g Re_s}\left[1 + 0.15\left(\alpha_g Re_s\right)^{0.687}\right]$$

$$Re_s = \frac{\rho_g d_s\left|\vec{V}_s - \vec{V}_g\right|}{\mu_g}$$

---

3. Gidaspow: [25]

$$K_{sg} = \tfrac{3}{4}C_D \frac{\alpha_s\alpha_g\rho_g\left|\vec{v}_s - \vec{v}_g\right|}{d_s}\alpha_g^{-2.65} \quad For \; \alpha_g > 0.8,$$

$$K_{sg} = 150\frac{\alpha_s(1-\alpha_g)\mu_g}{\alpha_g d_s^2} + 1.75\frac{\alpha_s\rho_g\left|\vec{v}_s - \vec{v}_g\right|}{d_s} \quad For \; \alpha_g \le 0.8,$$

where

$$C_D = \frac{24}{\alpha_l Re_s}\left[1 + 0.15(\alpha_l Re_s)^{0.687}\right]$$

$$Re_s = \frac{\rho_g d_s\left|\vec{V}_s - \vec{V}_g\right|}{\mu_g}$$

---

4. Huilin-Gidaspow: [33]

$$K_{sg} = \psi K_{sg-Ergun} + (1 - \psi)K_{sg-Wen\&Yu}$$

Stitching function

$$\Psi = \tfrac{1}{2} + \frac{arctan(262.5(\alpha_s - 0.2))}{\pi}$$

---

5. Gibilaro: [34]

$$K_{sg} = \frac{\left(\frac{18}{Re} + 0.33\right)\left(\rho_g|v_s - v_l|\right)}{d_p}\alpha_s\alpha_l^{-1.8}$$

---

6. Helland: [35]

$$K_{sg} = \tfrac{3}{4}C_D \frac{\alpha_s\alpha_g\rho_g\left|\vec{v}_s - \vec{v}_g\right|}{d_s}$$

where

$$C_D = C_{D,0} * C_{D,correction}$$

$$C_{D,0} = \frac{24}{\alpha_l Re_s}\left[1 + 0.15(\alpha_l Re_s)^{0.687}\right]$$

$$Re_s = \frac{\rho_g d_s\left|\vec{V}_s - \vec{V}_g\right|}{\mu_g}$$

$$C_{D,correction} = 924.8 - 1891.8 * \alpha_g + 968\alpha_g^2 \; for \; \alpha_g \ge 0.95$$

$$C_{D,correction} = \alpha_g^{-4.7} \; for \; \alpha_g \le 0.95$$

---

4. Result and Discussion

Before comparing drag model simulation results against experimental data, riser outlet mass flow rate transient profile as function of time for the six different drag models were shown in Figure 3. Riser outlet mass flow rate data for all investigated drag models were plotted for about 20 s in which all simulations have achieved a pseudo-steady state. It is interesting to note that the outlet mass flow rate profile for the Syamlal-O'Brien [26], Gidaspow [25], Wen–Yu [24] and Huilin-Gidaspow [33] drag models were fluctuating around the average mass flow rate of 1.04 kg/s compared to the Gibilaro [34] and Helland [35] drag models. Such comparison shows a distinct mass flow rate profile for different drag models; it can also be helpful to interpret model prediction behavior by analyzing the outlet mass flow rate.

Figure 4 shows the time-averaged solid volume fraction distribution for Syamlal-O'Brien, Gidaspow, Wen and Yu, Huilin-Gidaspow, Gibilaro and Helland drag models at a height of 1.5 and 2.0 m. For the Gibilaro and Helland drag model, the solid volume fraction transition from wall to core is sharp, whereas flow structure prediction with other drag model shows relatively uniform flow structure. In terms of quantitative comparison, the average mean solid volume fraction is indicated in the corresponding drag model name. The quantitative comparison reveals that the solid volume fractions are higher for Gibilaro and Helland than other drag models. When we compare drag model predictions for two different height, i.e., for 1.5 and 2.0 m, it observed that the solid volume fraction profile and average mean velocity value are approximately similar.

The influence of drag model is further illustrated by the time-averaged gas and solid phase axial velocity profile at a riser height of 1.5 m. The predicted time-averaged mean velocity for six different drag models were shown in Figure 5. It can be seen that the riser characteristic core–annulus flow structure observed in all cases. The highest gas velocity observed in the center and lowest value near the wall. Comparison clearly shows that the gas and the solid velocity magnitude and flow pattern variation for different drag models. For quantitative comparison, mean velocity value along the radial position were averaged and indicated in corresponding drag models. According to the value, gas phase axial average velocity is lowered by 4% from Syamlal-O'Brien, Gidaspow, Wen-Yu and Huilin-Gidaspow drag model to Gibilaro model and Helland drag model, whereas solid phase axial average velocity is lowered 39%. This over prediction of drag coefficient value results in higher mean solid velocity profile.

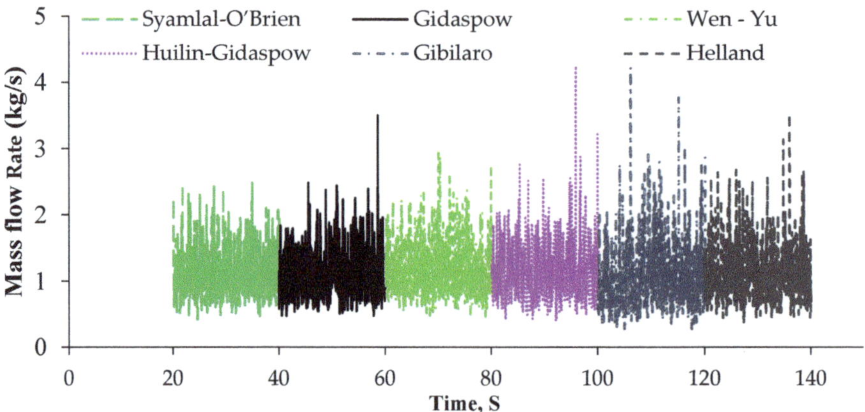

Figure 3. Influence of drag models on the outlet mass flow rate.

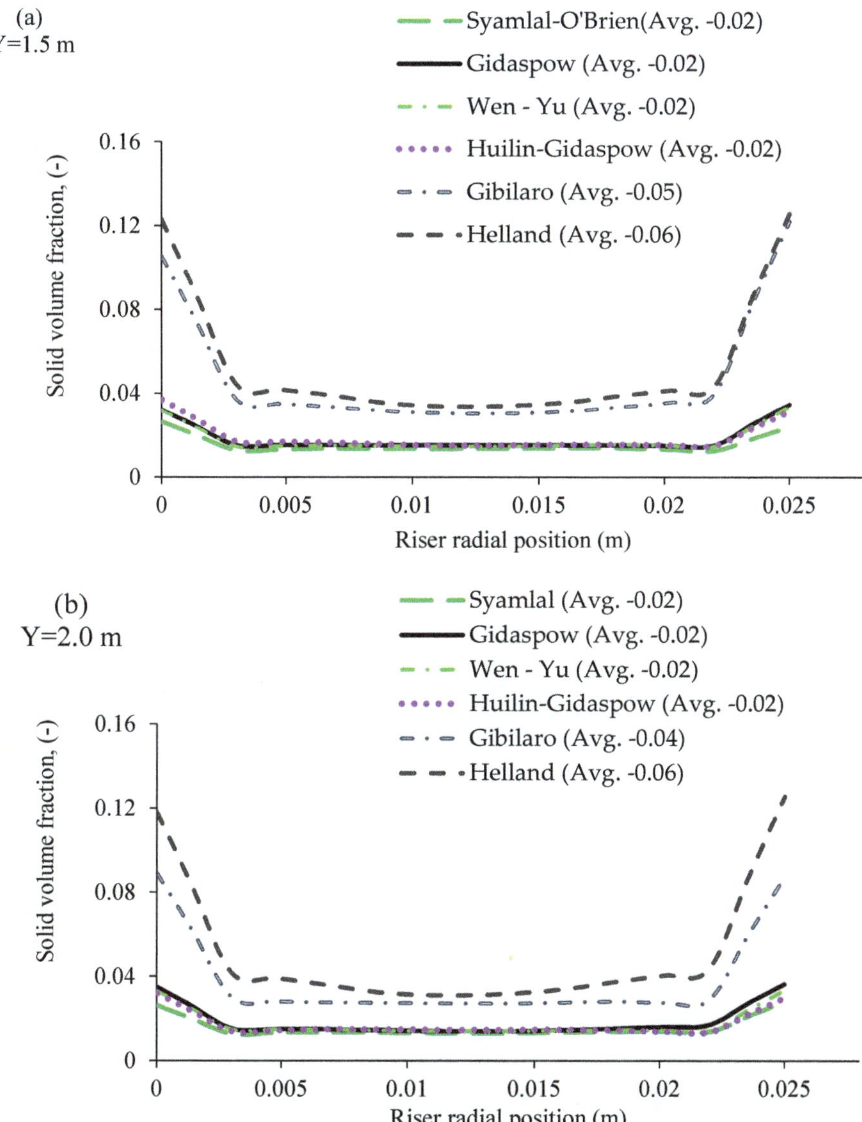

Figure 4. Time-averaged radial solid volume fraction for Syamlal-O'Brien, Gidaspow, Wen-Yu, Huilin-Gidaspow, Gibilaro and Helland drag models at height of (**a**) 1.5 and (**b**) 2 m.

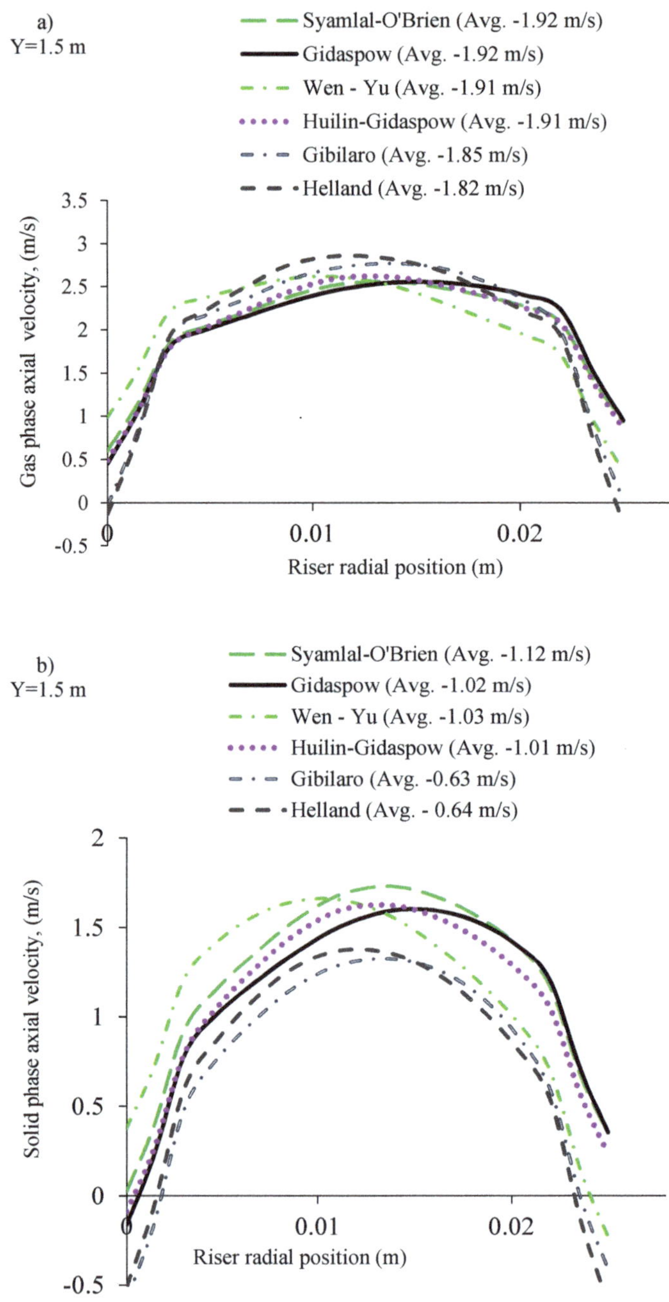

Figure 5. Influence of drag models (**a**) gas phase axial velocity (**b**) solid phase axial velocity at H = 1.5 m.

Figure 6 shows the time-averaged axial solid holdup profiles from different drag models. CFB riser experimental data exhibited L-shaped solid holdup profile. Syamlal-O'Brien, Gidaspow, Wen-Yu and Huilin-Gidaspow drag model predicts similar axial solid holdup and able to reproduce typical

L-shaped flow characteristics profile. Compare to other drag models, Gibilaro and the Helland drag model predicts higher solid holdup all along the height of riser. Interestingly solid holdup values being gradually declined from around 1.75 m to the outlet of the riser. The solid holdup prediction from the Syamlal-O'Brien, Gidaspow, Wen-Yu and Huilin-Gidaspow drag models were closer to the experimental values in the mid-section and upper section of the riser, whereas the Gibilaro and Helland drag model agreed reasonably well with the experimental data below 0.5 m. The overall performance of six different drag models was further assessed using root mean square error (RMSE) method as follows.

$$RMSE = \sqrt{\frac{1}{n}\sum_{i=1}^{n}\left(\varepsilon_s^{experiment} - \varepsilon_s^{CFD}\right)^2} \tag{1}$$

where $\varepsilon_s^{experiment}$ experimental solid holdup value and ε_s^{CFD} solid holdup predicted from six different drag models. Small RMSE indicates a closer model prediction to the experimental data. Figure 7 displays the RMSE for CFD simulation prediction with respect to six different drag models. RMSE values were calculated from different height of the riser and averaged. CFD simulation results with the Helland drag model shows highest RMSE value compared to that Syamlal-O'Brien drag model has the lowest value. This comparison demonstrates that the Syamlal-O'Brien, Gidaspow and Wen-Yu drag models predict closer to the experimental data. Similarly, we found Gibilaro drag models RMSE value is higher than other drag model. Comparison between experiment and simulated axial solid holdup value using RMSE data demonstrates that the simulation prediction all along the height of riser reflected on RMSE value. Among all test drag models, none of the drag model successfully able to fully capture the upper dilute and bottom dense region. Therefore, in next part of this study is to formulate drag reduction correlation for given gas velocity, solid circulation rate and material type using three-dimensional (3D) CFD simulation.

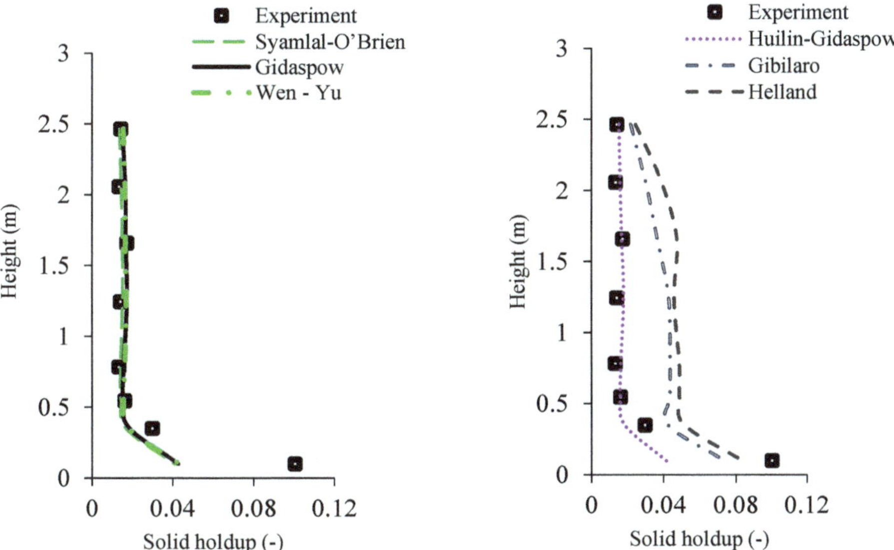

Figure 6. Influence of drag models on the axial solid holdup profile.

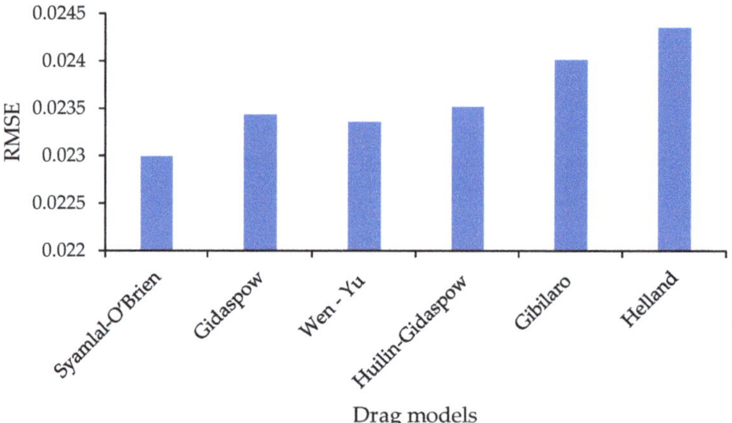

Figure 7. Root mean square error (RMSE) for six different drag model.

5. Conclusions

In this work, we investigate the limitation of six different drag models to predict the solid holdup profile in the riser section of circulating fluidized bed (CFB) reactor. A classical two-fluid model (TFM) was used to simulate 2D computational domain in ANSYS Fluent 2020R1 software. The six different simulation results were compared against solid holdup data along the height of CFB riser. Comparison analysis shows that axial solid holdup predicted by the Syamlal-O'Brien [26], Gidaspow [25], Wen-Yu [24] and Huilin-Gidaspow [33] models are the closest to the experimental data above the height of 0.5 m. However, they underpredict the bottom solid holdup compared to the Helland and Gibilaro drag model. Major prediction differences were observed seen in the bottom section of the riser with Syamlal-O'Brien, Gidaspow, Wen-Yu and Huilin-Gidaspow drag models. A discrepancy between the six different simulation prediction was quantify using the root-mean-square error (RMSE) calculation; the Syamlal-O'Brien [26] drag model was found to be the most accurate, followed by the Wen-Yu and Gidaspow drag model. Overall, their performance shows the requirement of the incorporating modified drag coefficient for individual flow condition and material type to predict upper dilute and bottom dense region.

Author Contributions: Conceptualization, M.U. and H.L.; Data curation, M.U., A.K., H.K. and D.L.; Formal analysis, M.U.; Funding acquisition, H.L.; Investigation, M.U., H.L., A.K., H.K. and D.L.; Methodology, M.U.; Project administration, H.L.; Software, M.U., A.K. and H.K.; Supervision, H.K. and M.U.; Validation, M.U., A.K. and H.L.; Visualization, M.U., A.K., H.K., D.L. and H.L.; Writing—original draft, M.U.; Writing—review & editing, M.U. and H.L. All authors have read and agreed to the published version of the manuscript.

Funding: This research received no external funding.

Acknowledgments: This work was supported by the Korea Institute of Energy Technology Evaluation and Planning (KETEP) granted financial resource from the Ministry of Trade, Industry & Energy, Korea (No. 20173030041290).

Conflicts of Interest: The authors declare no conflict of interest.

Nomenclature

C_D	Dimensionless drag coefficient
d_p	Solid particle mean diameter (μm)
D	CFB riser diameter (m)
e_{ss}	Particle–particle restitution coefficient
e_w	Particle–wall restitution coefficient
G_s	Solid flux (kg/m^{2}s)
g	Gravitational acceleration (m/s^2)
$g_{0,ss}$	Radial distribution function
H	CFB riser height (m)
$k_{\theta s}$	Diffusion coefficient for granular energy (kg/m s)
p_i	Pressure (Pa)
q_s	Granular temperature flux at the wall
Re_i	Reynolds number
U_g	Superficial gas velocity (m/s)
$\overrightarrow{U_s}$	particle slip velocity parallel to the wall
v_i	Velocity (m/s)

Greek Symbols

α_i	Phase i, volume fraction
$\alpha_{s,max}$	Solid volume fraction at maximum packing
$\gamma_{\theta s}$	Collisional dissipation of energy (kg/m^{3}s)
θ_i	Granular temperature (m^2/s^2)
λ_s	Solid phase bulk viscosity (kg/s/m)
μ_i	Shear viscosity (kg/s/m)
ρ_i	Phase i, density (kg/m^3)
τ_i	Stress tensor for phase i, (Pa)
K_{gs}	Gas–solid phase interphase momentum exchange coefficient, (kg/m^{3}s)
φ	Specularity coefficient
Φ_{gs}	Transfer rate of energy (kg/m^{3}s)

Subscripts

col	collisional
fr	frictional
g	gas phase
kin	kinetic
s	solid phase
ss	solid-solid
w	wall

Abbreviation

2D	Two-dimensional
CFD	Computational fluid dynamics
CFB	Circulating fluidized bed
SIMPLE	semi-implicit method for pressure-linked equations
TFM	Two-fluid model
RMSE	Root mean square error

Appendix A

Table A1. Gas–solid drag models used in literature.

References	Reactor Size H–Height (m) D–Diameter (m)	Flow Conditions U_g–Superficial gas velocity (m/s) G_s–Solid circulation rate (kg/m²s)	Solid Material Properties and Type ρ_s–Particle Density (kg/m³) d_s–Mean Particle Diameter (μm)	Drag Models Used (Comments)
Upadhyay and Park (2015) [16]	H: 3 m D: 0.025 m	U_g–2.0 m/s G_s – 39.15, 51.05, 73.21 kg/m²s	Silica sand ρ_s–2525 kg/m³ d_s–130 μm	Gidaspow drag [16]
Almuttahar and Taghipour (2008) [37]	H: 6.1 m D: 0.076 m	U_g–8, 6, 4, 8, 8, 4 m/s (Case 1~6) G_s–455, 355, 325, 254, 555, 94 kg/m²s (Case 1~6)	FCC particle ρ_s–1600 kg/m³ d_s–70 μm	Syamlal-O'Brien [17]
Almuttahar and Taghipour (2008) [38]	H: 6.1 m D: 0.0762 m	U_g–8.0 m/s G_s–453 kg/m³	FCC particle ρ_s–1600 kg/m³ d_s–70 μm	Syamlal-O'Brien [17]
Neri and Gidaspow, (2000) [39]	H: 6.58 m D: 0.075 m	U_g–2.61 m/s G_s–20.4 kg/m²s	FCC particle ρ_s–1654 kg/m³ d_s–75 μm	Gidaspow drag [17]
Chalermsinsuwan et al. (2009) [40]	H: 14.2 m D: 0.2 m	U_g–5.2 m/s G_s–489 kg/m²s U_g–3.25 m/s G_s–98.80 kg/m²s	FCC ρ_s–1712 kg/m³ d_s–76 μm	EMMS
Jin et al. (2010) [36]	H: 5.12 m D: 0.06 m	U_g–9.8, 8.6, 10.7 m/s G_s–546, 364 kg/m²s	Geldart B ρ_s–2580 kg/m³ d_s–385 μm	Syamlal-O'Biren [17]
Wilde et al. (2003) [41]	H: 14.434 m D: 1.56 m	U_g–3.36 m/s G_s–2.6 kg/m²s	Geldart A ρ_s–1550 kg/m³ d_s–60 μm	Gidaspow [16]

Table A1. *Cont.*

References	Reactor Size H-Height (m) D-Diameter (m)	Flow Conditions U_g–Superficial gas velocity (m/s) G_s–Solid circulation rate (kg/m²s)	Solid Material Properties and Type ρ_s–Particle Density (kg/m³) d_s–Mean Particle Diameter (μm)	Drag Models Used (Comments)
Koksal and Hamdullahpur (2005) [42]	H: 7.6 m D: 0.23 m	FCC U_g–3,5 m/s G_s–18,33 kg/m²s Silica sand U_g–5 m/s G_s–8 kg/m²s	Silica sand (Geldart B) ρ_s–2650 kg/m³ d_s–250 μm FCC ρ_s–1600 kg/m³ d_s–60 μm	Wen-Yu [15]
Shah et al. (2011) [28]	Low solid flux: H: 10.5 m D:0.09 m High solid flux: H: 14.2 m D: 0.2 m	Low solid flux: U_g–1.52 m/s G_s–14.3 kg/m²s High solid flux: U_g–5.2 m/s G_s–489 kg/m²s	FCC Low solid flux: ρ_s–930 kg/m³ d_s–54 μm High solid flux: ρ_s–1712 kg/m³ d_s–76μm	EMMS (Best prediction) Gidaspow [16]
Benyahia et al. (2005) [43]	H: 1.42 m D: 0.0142 m	U_g–14.85 m/s G_s–15 kg/m²s	Glass bead ρ_s–2500 kg/m³ d_s–70 μm	Syamlal O'Brien [17]
Vaishali et al. (2007) [29]	H: 7.9 m D: 0.152 m	Fast Fluidization (FF) U_g–3.2 m/s G_s–26.6 kg/m²s Dilute Phase Transport (DPT) U_g–3.9, 4.5 m/s G_s–33.7, 36.8 kg/m²s	Scandium coated with Parylene polymer ρ_s–2550 kg/m³ d_s–150 μm	Wen-Yu [15] Syamlal-O'Brien [17] (Best prediction)
Cloete et al. (2011) [44]	Periodic section: H: 0.8 m D: 0.076 m	U_g–3.5 m/s G_s–100 kg/m²s	FCC ρ_s–1500 kg/m³ d_s–67 μm	Syamlal-O'Brien [17]

Table A1. *Cont.*

References	Reactor Size H-Height (m) D-Diameter (m)	Flow Conditions U_g–Superficial gas velocity (m/s) G_s–Solid circulation rate (kg/m^{2}s)	Solid Material Properties and Type ρ_s–Particle Density (kg/m^3) d_s–Mean Particle Diameter (µm)	Drag Models Used (Comments)
Li et al. (2020) [45]	Transport section: H: 3 m D: 0.51 m Enlarged section: H: 0.152 m D: 0.635 m	U_g–1.6 m/s G_s–4.64 kg/m^{2}s	Iron-based oxygen carriers (Geldart D) ρ_s–2500 kg/m^3 d_s–1500 µm	Gidaspow [16]
Li et al. (2014) [46]	Square riser section: H: 9.14 m Cross sectional dimension: 0.146 * 0.146 m Circular NETL B22 CFB riser: H: 16.8 m D: 0.305 m Circular Malcus et al.'s CFB riser: H: 7 m D: 0.14 m	Square riser section: U_g–5.5 m/s G_s–40 kg/m^{2}s Circular NETL B22 CFB riser: U_g–5.14 m/s G_s–9.26, 40 kg/m^{2}s U_g–7.58 m/s G_s–14 kg/s Circular Malcus et al.'s CFB riser: U_g–4.7 m/s G_s–302 kg/m^{2}s	Square riser section: Sand particle ρ_s–2640 kg/m^3 d_s–213 µm Circular NETL B22 CFB riser: Glass beads ρ_s–2425 kg/m^3 d_s–59 µm High-density polyethylene (HDPE) beads ρ_s–863 kg/m^3 d_s–800 µm Circular Malcus et al.'s CFB riser: FCC ρ_s–1740 kg/m^3 d_s–89 µm	Gidaspow [16]
Wang et al. (2010) [30]	H: 10 m D: 0.076 m	U_g–8 m/s G_s–300, 500 kg/m^{2}s	Geldart A ρ_s–1500 kg/m^3 d_s–67 µm	Syamlal-O'Brien [17] (Best prediction) Wen-Yu [15] Gidaspow model

Table A1. *Cont.*

References	Reactor Size H-Height (m) D-Diameter (m)	Flow Conditions U_g–Superficial gas velocity (m/s) G_s–Solid circulation rate (kg/m²s)	Solid Material Properties and Type ρ_s–Particle Density (kg/m³) d_s–Mean Particle Diameter (μm)	Drag Models Used (Comments)
Zhang et al. (2015) [31]	H: 17 m D: 0.102 m	Case-1: U_g–8.6 m/s G_s–530 kg/m²s Case-2: U_g–8.6 m/s G_s–530 kg/m²s	Geldart B ρ_s–2951 kg/m³ d_s–89 μm	EMMS/Matrix (Best prediction) Gidaspow model
Zhang et al. (2015) [32]	H: 18.3 m D: 0.1 m	U_g–8.6, 8.0, 4.0 m/s G_s–171, 627, 635, 869, 209, 823, 857, 1045 kg/m²s	Geldart B Case-1: ρ_s–2951 kg/m³ d_s–89 μm Case-2: ρ_s–2620 kg/m³ d_s–154 μm Case-3 ρ_s–2620 kg/m³ d_s–154 μm	EMMS (Best prediction) Gidaspow [19]

B.

B.1. Continuity Equation and Momentum Conservation Equation: i = gas phase (g), j = solid phase (s)

- Continuity equation:

$$\frac{\partial}{\partial t}\left(\alpha_{i,j}\rho_{i,j}\right) + \nabla \cdot \left(\alpha_{i,j}\rho_{i,j}\vec{v}_{i,j}\right) = 0 \tag{A1}$$

B.2. Momentum Conservation Equation

- Momentum equation:

$$\frac{\partial}{\partial t}\left(\alpha_{i,j}\rho_{i,j}\vec{v}_{i,j}\right) + \nabla \cdot \left(\alpha_{i,j}\rho_{,ji}\vec{v}_{i,j}\vec{v}_{i,j}\right) = -\alpha_{i,j}\nabla p + \nabla \cdot \overline{\overline{\tau_{i,j}}} + \alpha_{i,j}\rho_{i,j}g + K_{i-j}\left(\vec{v}_i - \vec{v}_j\right) \tag{A2}$$

Solid and gas phase stress tensors:

$$\overline{\overline{\tau_s}} = \alpha_s\mu_s(\nabla\vec{v}_s + \nabla\vec{v}_s^T) + \alpha_s\left(\lambda_s - \frac{2}{3}\mu_s\right)\nabla.\vec{v}_s\overline{\overline{I}}$$

$$(\overline{\overline{\tau_g}} = \alpha_g\mu_g\left\{(\nabla\vec{v}_g + \nabla\vec{v}_g^T) - \frac{2}{3}\nabla.\vec{v}_g\overline{\overline{I}}\right\}$$

B.3. Kinetic Theory of Granular Flow (KTGF)

- Solid phase granular temperature equation:

$$\frac{3}{2}[\frac{\partial}{\partial t}(\rho_s\alpha_s\theta_s) + \nabla \cdot (\rho_s\alpha_s\vec{v}_s\theta_s)] = (-p_s\overline{\overline{I}} + \overline{\overline{\tau_s}}) : \nabla\vec{v}_s + \nabla \cdot (k_{\theta_s}\nabla\theta_s) - \gamma_{\theta s} + \phi_{sg} \tag{A3}$$

$$(-p_s\overline{\overline{I}} + \overline{\overline{\tau_s}}) : \nabla\vec{v}_s = \text{Generation of energy by the solid phase stress tensor}$$

Collisional energy dissipation:

$$\gamma_{\theta s} = \frac{12(1 - e_{ss}^2)g_{0,ss}}{d_p \sqrt{\pi}}\rho_s\varepsilon_s\theta_s^{3/2}$$

Diffusion coefficient for the granular energy:

$$k_{\theta_s} = \frac{150d_s\rho_s \sqrt{\theta_s\pi}}{384(1 + e_{ss})g_{0,ss}}\left[1 + \frac{6}{5}\alpha_s g_{0,ss}(1 + e_{ss})\right] + 2\rho_s\alpha_s^2 d_s(1 + e_{ss})g_{0,ss}\sqrt{\frac{\theta_s}{\pi}} \tag{A4}$$

where, $\phi_{sg} = -3K_{gs}\theta_s$ energy exchange between the gas and solid phase

- Solid phase pressure:

$$P_s = \alpha_s\rho_s\theta_s + 2\rho_s(1 + e_{ss})\alpha_s^2 g_{0,ss}\theta_s \tag{A5}$$

where, $g_{0,ss}$ radial distribution function

$$g_{0,ss} = \left[1 - \left(\frac{\alpha_s}{\alpha_{s,max}}\right)^{\frac{1}{3}}\right]^{-1}$$

- Solid bulk viscosity:

$$\lambda_s = \frac{4}{3}\alpha_s\rho_s d_s g_{0,ss}(1 + e_{ss})\sqrt{\frac{\theta_s}{\pi}} \tag{A6}$$

- Solid phase shear viscosity:

$$\mu_s = \mu_{s,col} + \mu_{s,kin} + \mu_{s,fr} \tag{A7}$$

$$\mu_{s,col} = \frac{4}{5}\alpha_s\rho_s d_s g_{0,ss}(1 + e_{ss})\sqrt{\frac{\theta_s}{\pi}} \tag{A8}$$

$$\mu_{s,kin} = \frac{\alpha_s d_s \rho_s \sqrt{\theta_s \pi}}{6(3 - e_{ss})}\left[1 + \frac{2}{5}(1 + e_{ss})(3e_{ss} - 1)\alpha_s g_{0,ss}\right] \tag{A9}$$

$$\mu_{s,kin} = \frac{10 d_s \rho_s \sqrt{\theta_s \pi}}{96\alpha_s(1 + e_{ss})g_{0,ss}}\left[1 + \frac{4}{5}g_{0,ss}\alpha_s(1 + e_{ss})\right]^2 \tag{A10}$$

B.4. Shear Stress and Collision Energy at the Wall

$$\tau_s = -\frac{\sqrt{3}\pi \rho_s \alpha_s g_{0,ss}\varphi \sqrt{\theta_s}}{6\alpha_{s,max}}\vec{U}_s \tag{A11}$$

$$q_s = \frac{\pi}{6}\sqrt{3}\varphi\frac{\alpha_s}{\alpha_{s,max}}\rho_s g_{0,ss}\sqrt{\theta_s}v_{s,w}^2 - \frac{\pi}{4}\sqrt{3}\frac{\alpha_s}{\alpha_{s,max}}(1 - e_w^2)\rho_s g_{0,ss}\theta_s^{3/2} \tag{A12}$$

References

1. Kunii, D.; Levenspiel, O. *Fluidization Engineering*, 2nd ed.; Butterworth-Heinemann Inc.: Boston, MA, USA, 1991.
2. Berruti, F.; Pugsley, T.S.; Godfroy, L.; Chaouki, J.; Patience, G.S. Hydrodynamics of circulating fluidized bed risers: A review. *Can. J. Chem. Eng.* **1995**, *73*, 579–602. [CrossRef]
3. Knowlton, T.M.; Grace, J.R.; Avidan, A.A. *Circulating Fluidized Beds*; Blackie Academic & Professional: London, UK, 1997.
4. Naren, P.R.; Lali, A.M.; Ranade, V.V. Evaluating EMMS model for simulating high solid flux risers. *Chem. Eng. Res. Des.* **2007**, *85*, 1188–1202. [CrossRef]
5. Abgrall, R. On essentially non-oscillatory schemes on unstructured meshes: Analysis and implementation. *J. Comput. Phys.* **1994**, *114*, 45–58. [CrossRef]
6. Durlofsky, L.J.; Engquist, B.; Osher, S. Triangle based adaptive stencils for the solution of hyperbolic conservation laws. *J. Comput. Phys.* **1992**, *98*, 64–73. [CrossRef]
7. Liu, X.D.; Osher, S.; Chan, T. Weighted essentially non-oscillatory schemes. *J. Comput. Phys.* **1994**, *115*, 200–212. [CrossRef]
8. Cockburn, B.; Shu, C.W. Runge–Kutta discontinuous galerkin methods for convection-dominated problems. *J. Comput. Phys.* **2001**, *16*, 173–261.
9. Liu, Y.; Zhang, W.; Jiang, Y.; Ye, Z. A high-order finite volume method on unstructured grids using RBF reconstruction. *Comput. Math. Appl.* **2016**, *72*, 1096–1117. [CrossRef]
10. Liu, H.; Xu, K.; Zhu, T.; Ye, W. Multiple temperature kinetic model and its applications to micro-scale gas flows. *Comput. Fluids* **2012**, *67*, 115–122. [CrossRef]
11. Zhu, T.; Ye, W. Theoretical and numerical studies of noncontinuum gas-phase heat conduction in micro/nano devices. *Numer. Heat Tranf. B-Fundam.* **2010**, *57*, 203–226. [CrossRef]
12. Karpinska, A.M.; Bridgeman, J. CFD-aided modelling of activated sludge systems—A critical review. *Water Res.* **2016**, *88*, 861–879. [CrossRef]
13. Wang, J. Continuum theory for dense gas-solid flow: A state-of-the-art review. *Chem. Eng. Sci.* **2019**, *215*, 115428. [CrossRef]
14. Ding, J.; Gidaspow, D. A bubbling fluidization model using kinetic theory of granular flow. *AICHE J.* **1990**, *36*, 523–538. [CrossRef]
15. Kuipers, J.A.M.; Van Duin, K.J.; Van Beckum, F.P.H.; Van Swaaij, W.P.M. A numerical model of gas-fluidized beds. *Chem. Eng. Sci.* **1992**, *47*, 1913–1924. [CrossRef]
16. Upadhyay, M.; Park, J.H. CFD simulation via conventional Two-Fluid Model of a circulating fluidized bed riser: Influence of models and model parameters on hydrodynamic behavior. *Powder Technol.* **2015**, *272*, 260–268. [CrossRef]
17. Tsuji, Y.; Kawaguchi, T.; Tanaka, T. Discrete particle simulation of two-dimensional fluidized bed. *Powder Technol.* **1993**, *77*, 79–87. [CrossRef]
18. Deen, N.G.; Annaland, M.V.S.; Van der Hoef, M.A.; Kuipers, J.A.M. Review of discrete particle modeling of fluidized beds. *Chem. Eng. Sci.* **2007**, *62*, 28–44. [CrossRef]
19. Andrews, M.J.; O'Rourke, P.J. The multiphase particle-in-cell (MP-PIC) method for dense particulate flows. *Int. J. Multiph. Flow* **1996**, *22*, 379–402. [CrossRef]

20. Upadhyay, M.; Park, H.C.; Hwang, J.G.; Choi, H.S.; Jang, H.N.; Seo, Y.C. Computational particle-fluid dynamics simulation of gas-solid flow in a circulating fluidized bed with air or O2/CO2 as fluidizing gas. *Powder Technol.* **2017**, *318*, 350–362. [CrossRef]

21. Shah, M.T.; Utikar, R.P.; Pareek, V.K.; Tade, M.O.; Evans, G.M. Effect of closure models on Eulerian–Eulerian gas–solid flow predictions in riser. *Powder Technol.* **2015**, *269*, 247–258. [CrossRef]

22. Ranade, V.V. *Computational Flow Modeling for Chemical Reactor Engineering*; Academic Press: London, UK, 2001; pp. 19–20.

23. Agrawal, K.; Loezos, P.N.; Syamlal, M.; Sundaresan, S. The role of meso-scale structures in rapid gas–solid flows. *J. Fluid Mech.* **2001**, *445*, 151–185. [CrossRef]

24. Wen, C.Y. Mechanics of fluidization. *Chem. Eng. Prog. Symp. Ser.* **1966**, *62*, 100–111.

25. Gidaspow, D.; Bezburuah, R.; Ding, J. Hydrodynamics of circulating fluidized beds: Kinetic theory approach (No. CONF-920502-1). In *Fluidization VII: Proceedings of the 7th Engineering Foundation Conference on Fluidization*; Engineering Foundation: Gold Coast, Australia, 1992; pp. 75–82.

26. Syamlal, M.; O'Brien, T.J. Computer simulation of bubbles in a fluidized bed. *AIChE Symp. Ser.* **1989**, *85*, 22–31.

27. Ergun, S. Fluid flow through packed columns. *Chem. Eng. Prog.* **1952**, *48*, 89–94.

28. Shah, M.T.; Utikar, R.P.; Tade, M.O.; Pareek, V.K. Hydrodynamics of an FCC riser using energy minimization multiscale drag model. *Chem. Eng. J.* **2011**, *168*, 812–821. [CrossRef]

29. Vaishali, S.; Roy, S.; Bhusarapu, S.; Al-Dahhan, M.H.; Dudukovic, M.P. Numerical simulation of gas–solid dynamics in a circulating fluidized-bed riser with Geldart group B particles. *Ind. Eng. Chem. Res.* **2007**, *46*, 8620–8628. [CrossRef]

30. Wang, X.; Jin, B.; Zhong, W.; Xiao, R. Modeling on the hydrodynamics of a high-flux circulating fluidized bed with Geldart Group A particles by kinetic theory of granular flow. *Energy Fuels* **2010**, *24*, 1242–1259. [CrossRef]

31. Zhang, Y.; Lei, F.; Wang, S.; Xiang, X.; Xiao, Y. A numerical study of gas–solid flow hydrodynamics in a riser under dense suspension upflow regime. *Powder Technol.* **2015**, *280*, 227–2383. [CrossRef]

32. Zhang, Y.; Lei, F.; Xiao, Y. The influence of pressure and temperature on gas-solid hydrodynamics for Geldart B particles in a high-density CFB riser. *Powder Technol.* **2018**, *327*, 17–28. [CrossRef]

33. Huilin, L.; Gidaspow, D. Hydrodynamics of binary fluidization in a riser: CFD simulation using two granular temperatures. *Chem. Eng. Sci.* **2003**, *58*, 3777–3792. [CrossRef]

34. Gibilaro, L.G.; Di Felice, R.; Waldram, S.P.; Foscolo, P.U. Generalized friction factor and drag coefficient correlations for fluid-particle interactions. *Chem. Eng. Sci.* **1985**, *40*, 1817–1823. [CrossRef]

35. Helland, E.; Bournot, H.; Occelli, R.; Tadrist, L. Drag reduction and cluster formation in a circulating fluidised bed. *Chem. Eng. Sci.* **2007**, *62*, 148–158. [CrossRef]

36. Jin, B.; Wang, X.; Zhong, W.; Tao, H.; Ren, B.; Xiao, R. Modeling on high-flux circulating fluidized bed with Geldart Group B particles by kinetic theory of granular flow. *Energy Fuels* **2010**, *24*, 3159–3172. [CrossRef]

37. Almuttahar, A.; Taghipour, F. Computational fluid dynamics of a circulating fluidized bed under various fluidization conditions. *Chem. Eng. Sci.* **2008**, *63*, 1696–1709. [CrossRef]

38. Almuttahar, A.; Taghipour, F. Computational fluid dynamics of high density circulating fluidized bed riser: Study of modeling parameters. *Powder Technol.* **2008**, *185*, 11–23. [CrossRef]

39. Neri, A.; Gidaspow, D. Riser hydrodynamics: Simulation using kinetic theory. *AIChE J.* **2000**, *46*, 52–67. [CrossRef]

40. Chalermsinsuwan, B.; Piumsomboon, P.; Gidaspow, D. Kinetic theory based computation of psri riser: Part I—Estimate of mass transfer coefficient. *Chem. Eng. Sci.* **2009**, *64*, 1195–1211. [CrossRef]

41. Juray, D.W.; Guy, B.M.; Geraldine, J.H. The effects of abrupt T-outlets in a riser: 3D simulation using the kinetic theory of granular flow. *Chem. Eng. Sci.* **2003**, *58*, 877–888.

42. Koksal, M.; Hamdullahpur, F. CFD simulation of the gas–solid flow in the riser of a circulating fluidized bed with secondary air injection. *Chem. Eng. Commun.* **2005**, *192*, 1151–1179. [CrossRef]

43. Benyahia, S.; Syamlal, M.; O'Brien, T.J. Evaluation of boundary conditions used to model dilute, turbulent gas/solids flows in a pipe. *Powder Technol.* **2005**, *156*, 62–72. [CrossRef]

44. Cloete, S.; Amini, S.; Johansen, S.T. A fine resolution parametric study on the numerical simulation of gas–solid flows in a periodic riser section. *Powder Technol.* **2011**, *205*, 103–111. [CrossRef]

45. Li, S.; Shen, Y. Numerical study of gas-solid flow behaviors in the air reactor of coal-direct chemical looping combustion with Geldart D particles. *Powder Technol.* **2020**, *361*, 74–86. [CrossRef]

46. Li, T.; Pannala, S.; Shahnam, M. CFD simulations of circulating fluidized bed risers, part II, evaluation of differences between 2D and 3D simulations. *Powder Technol.* **2014**, *254*, 115–124. [CrossRef]

Publisher's Note: MDPI stays neutral with regard to jurisdictional claims in published maps and institutional affiliations.

 *chemengineering*

Article

Volume of Fluid Computations of Gas Entrainment and Void Fraction for Plunging Liquid Jets to Aerate Wastewater

Ali Bahadar

Department of Chemical and Materials Engineering, King Abdulaziz University, Rabigh 21911, Saudi Arabia; absali@kau.edu.sa

Received: 19 July 2020; Accepted: 9 October 2020; Published: 18 October 2020

Abstract: Among various mechanisms for enhancing the interfacial area between gases and liquids, a vertical liquid jet striking a still liquid is considered an effective method. This method has vast industrial and environmental applications, where a significant application of this method is to aerate industrial effluents and wastewater treatment. Despite the huge interest and experimental and numerical efforts made by the academic and scientific community in this topic, there is still a need of further study to realize improved theoretical and computational schemes to narrow the gap between the measured and the computed entrained air. The present study is a numerical attempt to highlight the air being entrained by water jet when it intrudes into a still water surface in a tank by the application of a Volume of Fluid (VOF) scheme. The VOF scheme, along with a piecewise linear interface construction (PLIC) algorithm, is useful to follow the interface of the air and water bubbly plume and thus can provide an estimate of the volume fraction for the gas and the liquid. Dimensionless scaling derived from the Fronde number and Reynolds number along with geometric similarities due to the liquid jet's length and nozzle diameter have been incorporated to validate the experimental data on air entrainment, penetration and void fraction. The VOF simulations for void fraction and air-water mixing and air jet's penetration into the water were found more comparable to the measured values than those obtained using empirical and Euler-Euler methods. Although, small overestimates of air entrainment rate compared to the experiments have been found, however, VOF was found effective in reducing the gap between measurements and simulations.

Keywords: liquid plunging jet; waste water treatment; VOF; void fraction; air entrainment

---

1. Introduction

The significance of the liquid plunging jet's entrainment of a gas lies in achieving a large interfacial area between gas and liquid, which is useful for a large number of environmental and industrial applications. Fuel consumption optimization of ships is compromised due to the formation of a wake on the water surface associated to air being entrained across the boundary layer [1,2]. Pressurized water reactors (PWRs) face partially filled cold legs associated to emergency core cooling as a result of gas entrainment when the coolant strikes the water's surface [3,4]. Plunging jets have been proved efficient methods to aerate industrial effluents [5–8]. Other applications include mineral-processing flotation cells [9,10] and oxygenation of cell bioreactors [11]. In another industrial process involving the casting of polymers and glass [12], promoting enhanced interaction between gas and liquid aeration is significant for the efficient formation of products.

Among the several applications of the plunging jet mentioned above, aeration and floatation of industrial effluents and wastewater treatment [8,13] represent a huge application of this method. For aerobic treatment processes, such as the activated sludge process, plunging jet aeration systems provide a simple and inexpensive method of supplying oxygen for wastewater treatment [13,14].

Bin [13] has comprehensively reviewed gas entrainment by plunging liquid jets. Numerous studies are reported in his review on single plunging jets which reveal that gas entrainment by a plunging liquid jet is a complex process. The main quantities that determine the functioning of a plummeting jet, including the reduced critical entrainment's velocity, u_e, and the amount of entrained gas rate, Q_g, which depends on parameters expressed as D_j, u_j, ρ_g, ρ_l, g, μ_j, μ_g, v' σ, λ, respectively, where, D_j is the liquid jet's diameter, u_j is the liquid jet's velocity at the time of contact with the pool liquid, ρ_g is the gas density (i.e., air), ρ_l is the liquid's density of the jet and the liquid in the pool, g is the acceleration constant due to gravity, μ_j is the liquid viscosity (i.e., water), μ_g is the gas viscosity (i.e., air), v' is the geometry of the case has been approximated by refereeing to a characteristic length L.

The occurrence of air entrainment associated to the water jet impacting into a liquid pool has been found an efficient phenomenon to form larger interfacial areas between gas and liquid, for instance in environmental industries, plunging liquid jets can cause agitation when contacting the surface of a liquid pool to improve exchange between the gas and liquid [13,15,16]. For industrial effluent and water treatment, the influence due to the combined inertial as well as intense turbulence is to enhance the mass transfer of gases (i.e., nitrogen, oxygen, hydrogen sulphide, etc. and organic compounds). Air entrainment generally occurs due to two mechanisms that complement each other, one of these is the interfacial shear acting at the intermittent region between the water and the air, which shrinks the air boundary layer. However, the second phenomenon relates to the entrapment of the air in the vicinity of the plunging jet when striking a liquid pool [17]. Entrapment of air at the water surface causes a large scale cavity with the cylindrical shape at the bottom that is pinched off, thus forming a bubbly plume that moves downward below the impact surface of the water [18].

With a huge increase in computing power in the last few decades, several numerical models; two-fluid [19,20]; large eddy simulation (LES) [21,22], direct numerical simulation (DNS) [23], Euler-Euler method [22,24] and many others have been built to determine understanding of multiphase, gas and liquid plunging jets. Among these VOF, proposed by Hirt and Nichols [25], is useful to simulate the interface interaction between gas and liquid flows. Ma et al. [19] applied a Eulerian/Eulerian two-fluid computational multiphase fluid dynamics (CMFD) model along with a supplement sub-grid air entrainment model to predict the distribution of void fraction in vertical plunging water jets. Miwa et al. [4] made use of analytical and dimensionless numbers such as the Weber number and Laplace length scale to propose a predictive model for the computation of air entrainment ratio (entrained air rate/impinging jet rate), which deviated from the experimental values by about 15.9%. Yin et al. [20] used the volume of fluid and coupled level set method to investigate the numerical response of a vertical plunging jet. The computational values were in accordance with reported literature values on the subject matter. Different vertical jet parameters such as the air void fraction, liquid velocity field, and turbulent kinetic energy were analyzed by altering the length between the still water level and the nozzle exit. The velocity of the nozzle exit had no role in the size and shape of inpinging bubbles. A revised predictive equation relating the axial separation and the centerline velocity ratio was formulated with the help of data for the coupled level set and volume of fluid methods, and it exhibited a greater predictability capacity than the mixture and level set methods. A comparison was drawn between turbulent kinetic characteristics such as radial and vertical distribution and the highest value location to that of submerged jets.

Several studies have been performed to track the motion of the free surface [21,22], the distortion of the interface between liquid and gas in columns [23,24], and the VOF has been applied to visualize both the plunging of the liquid jet into the still liquid and the subsequent entrained bubble plume phenomena [25,26].

The present attempt aims to focus on air being entrained by a water jet striking a still water pool using the VOF gas-liquid model. The Navier-Stokes equations are solved using the CFD code ANSYS-Fluent. The objective of the numerical investigation is to validate the applicability and accuracy of the VOF model to replicate a circular water jet sinking into a still water pool. A VOF scheme along with PLIC proved effective to characterize the distinction between the gas and liquid phase, which can prove useful in the present case to determine the bubbly plume and void fraction along the plume. However, our main emphasis is to determine the distribution of air void fraction at various depths of the plume from the water-free surface and its corresponding impact on the values of the air entrainment rate.

2. Methodology

2.1. Physical Parameters

The schematic illustration in Figure 1 shows a vertical plunging water jet of diameter D_j, which intrudes into the still surface of a water tank at a velocity u_j. The relevant dimensionless numbers, that describe the physics of the phenomenon are the Reynolds number (Re), Froude number (Fr), and Weber number (We). Reynolds number can be expressed as:

$$Re = \frac{\rho_w u_j D_j}{\mu_w} \tag{1}$$

where, u_j is the jet's velocity when impacting the still surface of the water pool, D_j is the jet's diameter at the instance when the jet impacts the still surface of the water, ρ_w is the water density, i.e., 1000 kg/m^3 and μ_w is the water's viscosity, i.e., 1×10^{-3} pa.s. Froude number is expressed as:

$$Fr = \frac{u_j}{\sqrt{gD_j}} \tag{2}$$

where, g is the gravitational acceleration, i.e., 9.8 m^2/s. Weber number is written as:

$$We = \frac{\rho_w u_j^2 D_j}{\sigma} \tag{3}$$

where, σ is the surface tension corresponding to the air-water interface, 0.072 Nm^{-1}.

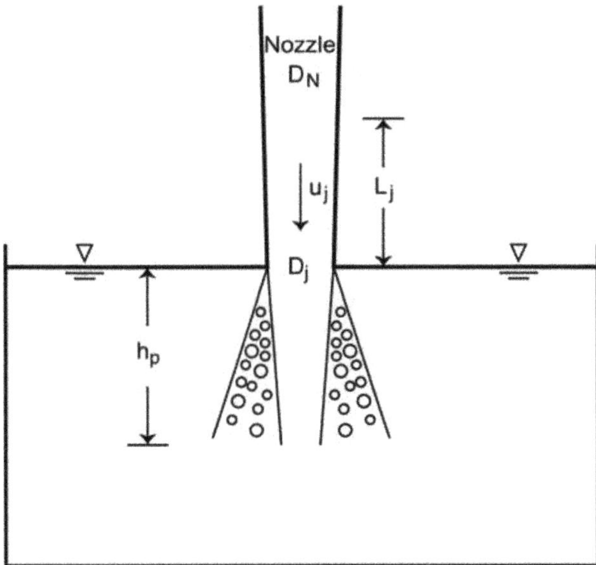

Figure 1. A schematic of a vertical plunging jet.

2.2. Numerical Scheme

The computational simulation has been conducted utilizing the commercial code Fluent applying VOF to characterize the multiphase flow. The VOF model, developed by Hirt and Nicholas [27], can track the interface between the two phases, so it can be used in all those studies where an interface between two phases exists. A two-equation, k-ε scheme using standard wall values in the vicinity of the wall layer has been applied to describe the turbulence.

The VOF has been applied here to solve multiphase flows. The VOF method deals with flow equations through averaging them in terms of volume to attain a group of equations, whereas the interface between the gas and liquid can be followed by applying a phase marker (α), better named volume function, which can be expressed as: (i) If $\alpha = 1$ (i.e., computational cell occupies phase 1, e.g., liquid), (ii) If $\alpha = 0$ (computational cell occupies phase 2, e.g., gas), (iii) $0 < \alpha < 1$ (both phases are present and an interface exists within the computational cell). For those cells containing either phase 1 or 2, the formulation consists of mass conservation equations and momentum conservation equations containing properties of the phase present in the cell. If both phases exist within the cell, then the properties of the air and water mixture are evaluated through a mean voidage of the fluids present within the respective cell, thus ensuring to acquire a single set of equations by averaging the variables, for the present case, these variables are density and viscosity, their averaging is expressed as:

$$\rho = \alpha\rho_l + (1 - \alpha)\rho_g \tag{4}$$

$$\mu = \alpha\mu_l + (1 - \alpha)\mu_g \tag{5}$$

Using averaging variables and considering that across the interface, the velocity corresponding to the two phases is uniform, the continuity and momentum equations [28,29] expressed as [30,31]

$$\frac{\partial \rho}{\partial t} + \nabla.(\rho u) = 0 \tag{6a}$$

$$\frac{\partial}{\partial t}\rho u + \nabla.(\rho u u) = -\nabla p + \nabla.(\mu \nabla u) + \rho g + F_S \tag{6b}$$

where, F_S symbolizes the continuum surface force (CSF) [30,32]. This force acts across the interface, containing the interfacial and subsequent surrounding cells. F_S can be expressed as:

$$F_S = \sigma \frac{\rho_l}{\frac{1}{2}(\rho_l + \rho_g)}\kappa\nabla\alpha \tag{7}$$

where κ represents the curve of the surface due to air or water on it. κ can be computed [31,32] as:

$$\kappa = -\nabla.\vec{n} = -\nabla.\frac{\nabla\alpha}{|\nabla\alpha|} \tag{8}$$

where, $\vec{n}$ is the unit vector representing the surface perpendicular to the interface, however, at the wall, the surface normal to the computational cell, adjacent to the wall will be considered [30,32,33], $\vec{n}$ is expressed as:

$$\vec{n} = \vec{n_x}cos\theta + \vec{n_y}sin\theta \tag{9}$$

where θ is the angle between the interface and the wall and $\vec{n_x}$ and $\vec{n_y}$ stand for the unit vectors along tangential and normal to the wall, respectively. However, this is to confirm that all simulations that have been obtained here, are at a contact angle of 90° from the wall.

Since the volume fraction, α, is a property of the fluid, which changes along with the fluid, thus, its evolution is dictated by the simple advection relation, expressed as [34]:

$$\frac{\partial \alpha}{\partial t} + \nabla.u\alpha = 0 \tag{10}$$

Due to the fact the VOF model is more flexible and efficient compared to others used to track liquid-gas interfaces, the VOF has become useful in examining a wide range of gas-liquid problems [21,35,36]. Progression of the interface as a function of time between the gas and liquid is tracked in the computational cell using scheme, piecewise linear interface construction (PLIC). As prescribed by Youngs [37], the method considers a linear form of the interface between the two phases within all those cells where the two phases meet each other. Linear behavior of the interface within a cell (Figure 2) and distributions of the normal and tangential velocity at the face have been used to compute the advection of fluid (Equation (10)) across each face of the cell. Subsequently, the void fraction in every cell (α) is determined owing to the fluxes, computed in the previous cell.

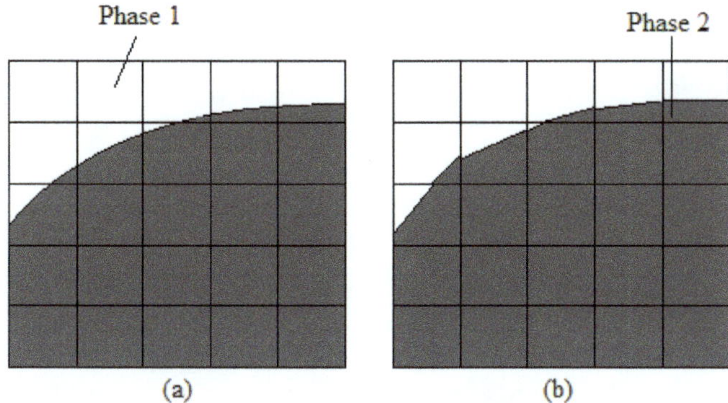

Figure 2. (**a**) Real Interface Boundary; (**b**) Linear Interface Reconstructed Within Each Cell Using PLIC, Following [31].

Turbulence Equations

The liquid leaving the nozzle has a velocity high enough to be in a turbulent regime. The gas entrapped due to the liquid jet interacting the surface of the still liquid pool and the gas-liquid two phase flow thus formed, plunge into the pool of liquid at a mixed velocity in the range of turbulent flow. Thus, it is reasonable to compute the gas-liquid jet flow in the liquid pool by application of equations for turbulent quantities acquired by the liquid and gas. The flow variables (e.g., velocity, pressure, etc.), contain mean and fluctuating components. Thus, incorporating the instantaneous equations (Equation (6a,b)) having decomposed variables and utilizing averaging of Reynolds stresses to the Navier-Stokes equations, along with the kinetic energy, k, of turbulence, and its dissipation rate, ε. After dropping the overbar on the mean variables for convenience, The Reynolds-averaged equation after ignoring the overbar for mean quantities can be expressed in a general form as:

$$\frac{\partial(\rho\varphi)}{\partial t} + div(\rho u\varphi) = div\left(\Gamma_\varphi grad\varphi\right) + S_\varphi \tag{11}$$

where φ represents the scalar variable, Γ_φ is the diffusion coefficient and S_φ is the source term. The representation of these quantities for different variables are given in Table 1. Also, the radial (r)

and axial (z) momentum equations along with k and ε equations and values of the respective constants appearing in these equations are also part of this Table.

$$\mu_t = C_\mu \rho k^2, \ \mu_{eff} = \mu + \mu_t, \ C_\mu = 0.09, \ C_1 = 1.44, \ C_2 = 1.92, \ \sigma_k = 1.0, \ \sigma_\varepsilon = 1.3$$

Transport equations relating to momentum are solved to determine turbulence quantities of the two-phase flow by sharing the variables k and ε as well as the Reynolds stresses (i.e., $-\rho u'u'$, $-\rho u'v'$, u' and v' are fluctuating velocities along with the axial and transverse flow) by the phases across the flow. However, in the governing equations for turbulent production (k) and its dissipation (ε), presented in Table 1, the consequence of turbulence (i.e., Reynolds stresses) is linked to the average flow through the standard k-ε turbulence model [38]. The standard version of the turbulence model has been chosen in our study here due to the sensible computing time and is useful for high Reynolds number flows.

Table 1. Source terms and values of coefficients of Generic Transport Equations [37].

Equation	φ	Γ_φ	S_φ
Continuity	φ/ρ	0	0
Radial momentum	v	μ_{eff}	$-\frac{\partial p}{\partial r} + \frac{\partial}{\partial z}\left(\mu_{eff}\frac{\partial u}{\partial r}\right) + \frac{1}{r}\frac{\partial}{\partial r}\left(r\mu_{eff}\frac{\partial v}{\partial r}\right) - \frac{2\mu_{eff}v}{r^2}$
Axial momentum	u	μ_{eff}	$-\frac{\partial p}{\partial z} + \frac{\partial}{\partial z}\left(\mu_{eff}\frac{\partial u}{\partial z}\right) + \frac{1}{r}\frac{\partial}{\partial r}\left(r\mu_{eff}\frac{\partial v}{\partial z}\right) + g$
k Equation	k	$\mu + \frac{\mu_l}{\sigma_k}$	$\mu_t\left[\left(\frac{\partial u}{\partial r} + \frac{\partial v}{\partial z}\right)^2 + 2\left(\frac{\partial u}{\partial z}\right)^2 + 2\left(\frac{\partial v}{\partial r}\right)^2 + 2\left(\frac{v}{r}\right)^2\right] - \rho\varepsilon$
ε Equation	ε	$\mu + \frac{\mu_l}{\sigma_k}$	$C_1\frac{\varepsilon}{k}\mu_t\left[\left(\frac{\partial u}{\partial r} + \frac{\partial v}{\partial z}\right)^2 + 2\left(\frac{\partial u}{\partial z}\right)^2 + 2\left(\frac{\partial v}{\partial r}\right)^2 + 2\left(\frac{v}{r}\right)^2\right] - C_2\rho\frac{\varepsilon^2}{k}$

2.3. Simulation Plan

Geometry, Mesh, and Boundary Conditions

The 2d meshed domain of the front section of the plunging liquid jet assembly is shown in Figure 3. The meshed domain is used to compute the liquid jet plunging into the pool of still water with subsequent gas entrainment as well as the development of the bubbly plume. The computational domain consists of a 0.15 m radius, where the height varies between 0.397, 0.42 and 0.47 m, which comprises the liquid pool's depth (0.32 m), nozzle pipe (50 mm), and liquid jet's length, with values of 27, 50 and 100 mm. The liquid jet's lengths adopted here are due to the nozzle sizes, i.e., 5, 6.83, 12.5 and 25 mm, which have been chosen here to follow Chanson's [39] experimental setup. The dimensions of the Chanson's experimental tank are sufficiently large so as not to have any impact on the contact of the water jet in the vicinity of the water-free surface. Here, the dimensions of the domain, including sizes of the water jet and the water pool (i.e., r = 0.15 m and h = 0.47 m) are sizeable enough that the boundaries have no influence, however, the jet's domain is much smaller than the pool dimensions. Due to the preference of 2d geometrical scheme and symmetry boundary conditions, the meshed region is reduced to about 25% of the entire geometry, which is useful to reduce the computing time as well as falling within the limitations of the computing grid storage.

The inward flowrate exiting the nozzle located at the top of the water free surface and the out rate of the geometry is located at the bottom, also, wall and outlet pressure constitute the boundary conditions [40]. Water flowing through the nozzle follows a constant mean velocity profile. At the wall, no slip has been assumed. Further, an open boundary condition is considered to model the top of the domain with gauge pressure 0 Pa. However, the outlet flow from the horizontal section of the tank should follow pressure outlet conditions.

Clustering is applied towards the centerline of the liquid jet and the top water free surface to track the phases owing to the liquid and the gas phases, along with the deformation of the interface between the gas and the liquid since the tiniest gas bubble experienced in the experiment involving plunging water jet utilizing tap water is nearly 0.7 mm [39]. Thus, the cell sizes being retained are of asize smaller than the tiniest bubbles within the plume.

The fluid domain can be divided into two regions: the region that has to inhibit the jet impingement as well as the interaction regions, appropriately referred here as region of interest (ROI) between the impinged jet and the surrounding water, the mesh size in the region of interest (ROI) may approximately be within 0.01–0.03 mm, whereas in the region that is external to the main flow region (ROI) the mesh size has been keep a little coarse which is around 0.7–0.85 mm. In this way we have tried to reduce the total number of meshes with the best-attained results for the parameters of our concern (i.e., jet impingement length and the jet diameter) with the optimized meshing.

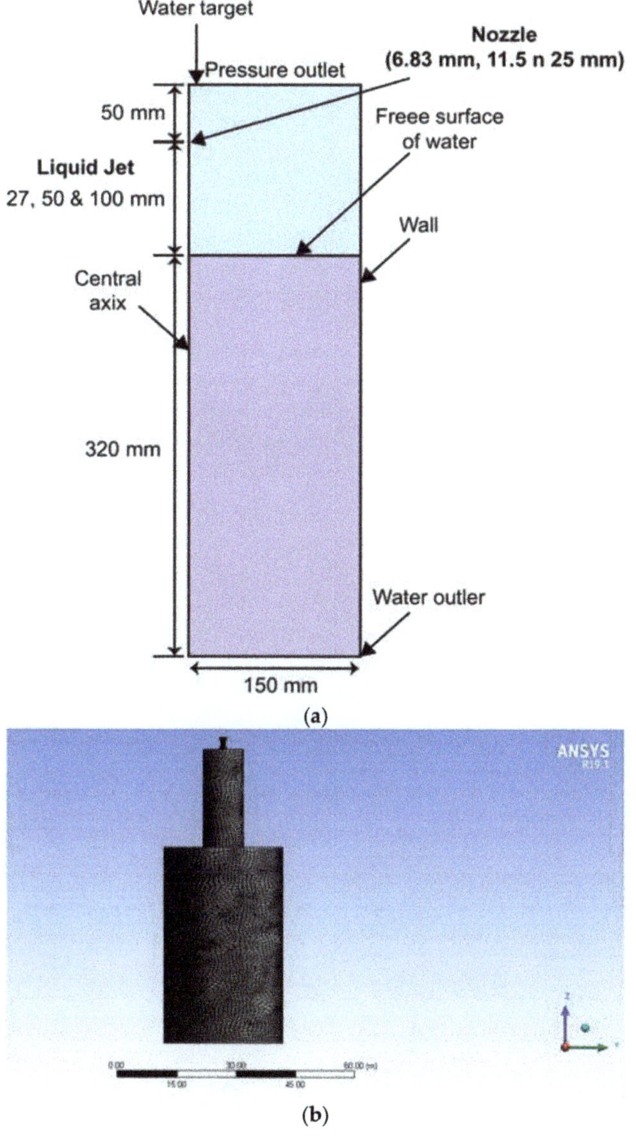

Figure 3. (a) Boundary Conditions and Dimensions; (b) Simulation Geometry.

Along the region surrounding the centerline, the cells are quadratic having excellent computational prospects owing to skewness factor and orthogonal quality [41]. At all faces of the computational cell, boundary values for physical and flow properties are specified, which include gas and liquid velocity components, acting pressure (which is atmospheric pressure in the present set-up), the level function, the gas number density, the turbulent kinetic energy, k, and the turbulent dissipation, ε. To model the liquid jet at the top surface of the domain, the vertical share of the jet's velocity is set as the nozzle outlet velocity, whereas, all remaining components of the velocity relating to the liquid and the gas as well as the function owing to the bubble's number density are considered as zero. As stated, before that for pressure, a zero-gradient boundary condition has been set, however, the function associated with the level set has been considered such that the periphery of the liquid's jet coincides with the level function having the zero, whereas, its value elsewhere by application of the level function is chosen to specify its zero value that coincides with that of the periphery of the liquid jet, whereas, its value elsewhere is equivalent to the distance from the level setting as zero. The model variables due to the turbulence, k, and ε, relating to their free-stream values as $9 \times 10^{-4}/\mathrm{Re}$ and 0.9, respectively [21] have been set, which, have been verified to be significantly small so as to not affect the simulations.

For computations relating to fluid volumes, the mass rate at outlet boundaries has been set as less than 10^{-6} and the 2-D derivatives were used in the conservation equations are discretized by the use of the quick scheme, while, geometrical re-construction has been applied to achieve discretization of the void fraction in advection Equation (10). The pressure-velocity coupling has been achieved through the algorithm, pressure-inherent with the splitting of operators, PISO [42], whereas, time discretization is used as the inherent first order function applicable to the conservation equations.

An extensive grid independence test was conducted by varying the mesh size with the same Hex type of mesh elements. The varying mesh size with the total number of meshes along with the its effect on the penetration length and jet diameter has been presented in Table 2.

Table 2. Grid Independence Test.

Case No.	Jet Velocity (m/s)	Mesh Size (Millions)	Element	Type	Mesh	Provision	Jet Penetration Length/Dia
01	1.7–15.7	0.31	Hex	Cooper	Fine	More fine in upper region near the jet impingement zone up till the middle of the rig	27.3–39.7/6.83–11.2
02	1.7–15.7	0.35	Hex	Cooper	Fine	More fine in upper region near the jet impingement zone up till the middle of the rig	31.8–41.2./7.9–13.2
03	1.7–15.7	0.37	Hex	Cooper	Fine	More fine in upper region near the jet impingement zone up till the middle of the rig	34.2–45.7/10.4–17.1
04	1.7–15.7	0.41	Hex	Cooper	Fine	More fine in upper region near the jet impingement zone up till the middle of the rig	37.88–49.12/12.7–20.41
05	1.7–15.7	0.43	Hex	Cooper	Fine	More fine in upper region near the jet impingement zone up till the middle of the rig	55.79–67–56/14.77–22.18
06	1.7–15.7	0.48	Hex	Cooper	Fine	More fine in upper region near the jet impingement zone up till the middle of the rig	74.58–89.18/18.79–24.87
07	1.7–15.7	0.57	Hex	Cooper	Fine	More fine in upper region near the jet impingement zone up till the middle of the rig	84.58–99.98/19.55–24.98
08	1.7–15.7	0.64	Hex	Cooper	Fine	More fine in upper region near the jet impingement zone up till the middle of the rig	85.58–99.99/19.55–24.99
09	1.7–15.7	0.69	Hex	Cooper	Fine	More fine in upper region near the jet impingement zone up till the middle of the rig	85.58–99.99/19.55–24.99
10	1.7–15.7	0.75	Hex	Cooper	Fine	More fine in upper region near the jet impingement zone up till the middle of the rig	85.58–99.99/19.55–24.99

Grid independence tests show that upon varying the density of the meshes from 0.31 million to 0.75 million, the highest values of the jet penetration length to the diameter of the jet has been found to be equal to its highest values i.e., (jet penetration length/jet dia 84.58–99.98/19.55–24.98). The highest values of the jet penetration length/jet dia has been achieved at 0.57 million meshes. After that computations were obtained for higher mesh sizes, however, the values for both quantities only differed by 1.0–0.01/0–0.01. Thus in order to save computational resources and time it was decided to continue with the 0.57 million meshes with the provision that a higher mesh resolution has to be used in the area where the jet impingement occurred all the way till the end of bubbly jet. As the total number of cases ranged to 9 for each mesh density thus we presented the cumulative results of the grid independence in the form of the ranges for the properties of interest i.e., the jet penetration length and the jet dia. in Table 2.

3. Numerical Results

To achieve similarity between the computations and the experimental results, non-dimensionized equations have been set to achieve right values for the jet's size and the jet velocity that lead to Reynold's numbers of 12,050–105,600 and Froude numbers ranging from 5 till about 9, which correspond to the experimental liquid jet velocity (1.79 m/s–4.4 m/s) analogous to the empirical nozzle dia (6.83, 12.5 and 25 mm). These values were considered for the purpose of comparison with the simulations relating to the void fraction. For evaluation of simulated outcomes for penetration of bubbly plume, experimental values [43] related to nozzle dia (0.3, 1.3, 2.4 mm) and jet velocity, ranging from 6.79–15.7 m/s, were used to evaluate the accuracy of the simulations. All these details including the geometrical dimensions and fluids properties along with the respective values of Re and Fr are summarized in Table 3.

Simulations have been performed initially iterating for a steady liquid phase alone. A steady-state solution has been achieved after time iterations of approximately 2000 with 0.015 dimensionless time, which is equivalent to real-time of nearly 0.1 ms. The steady conditions have been verified by inspecting no change in the time-averaged quantities by raising the number of iterations. Thus, before moving on to the simulations for the bubbly flow, a steady flow of the liquid phase alone has been set as the initial condition.

The simulation involving bubbly flow has been initiated by setting on the multiphase model. The bubbly flow simulation has been obtained with the time step the same as of the single phase, and bubbles travel vertically down across the computational domain towards the region where the bubbles' buoyancy is sufficient to enable the bubbles to move up. However, for simulating gas entrainment and void fraction profiles along with a depth of water pool at $2D_j$ from the contact surface, here air entrainment and void fraction measurements have been achieved experimentally. The bubbles have taken nearly 600 time periods to cover the computational domain equivalent to $2D_j$. The time-averaged void fractions have been obtained after attaining a steady value after approximately 3000 time periods. The measurements such as gas entrainment and void fraction, made by [39] involving water jet sinking into the water pool have been used for comparison with the simulations.

Table 3. (a) Geometry of Plunging Liquid Jet. (b) Operating Conditions and Fluid Properties.

(a)

SIM No.	Nozzle Dia (mm)-D_N	Jet Dia (mm)-D_J	Jet Length (mm)-L_J
1	25	24	100
2	12.5	12.5	50
3	6.83	6.83	27.3
4	0.3, 1.3, 2.4	0.3, 1.3, 2.4	

(b)

SIM No.	Nozzle Dia (mm)-D_N	Jet Dia (mm)-D_J	Jet Length (mm)–L_J	Jet Velocity (m/s)	Profile Locations (m)	Fluid Properties and Details	Fr_j	Re_j
1	25	24	100	3.5 4.1 4.4		Tap water, $\mu_w = 1.015 \times 10^{-3}$ Pa.s $\rho_w = 1000$ kg.m^{-3} $\sigma = 0.055$ N/m	7.2 8.4 9	82,760 96,950 104,040
2	12.5	12.5	50	2.42 3.04 3.46	$0.8D_J$, $1.2D_J$, $2D_J$		7.1 8.8 10	29,805 37,440 42,615
3	6.83	6.83	27.3	1.79 2.16 2.49			7.1 8.5 9.7	12,050 14,540 16,760
4	0.3, 1.3, 2.4	0.3, 1.3, 2.4		9.83–12.0, 1.86–15.7, 2.92–6.79	Penetration Depth		181.2–221.2, 16.47–139, 19–44.3	2949–3600, 2418–20410, 7008–16296

3.1. Gas Entrainment and Developing Bubbly Regime

Sequential illustrations of the simulations presented in Figure 4 show the formation of a gas plume within the liquid pool. VOF simulation exhibits entrainment of the air by the water jet, that has been driven deeper in the tank. The entrained gas bubbles have been advected by the downward flow to form a plume. The depth of the plume associates with the momentum of the water jet, however, bubbles can only be driven down to a depth where the bubble's buoyancy overcomes the momentum owned by the liquid jet, thus, bubbles rise due to the buoyancy to reach the surface of the pool, where they the gas separates the liquid. Similar computational results were found by [18,21,25]. The volumetric downward gas rate at a given level below the surface is given by:

$$Q_g(z) = \int_z q_g dS \tag{12}$$

where q_g is the volumetric rate flux of the gas. Since the 1980s, empirical predictive modeling equations relating to plunging liquid jets have been proposed [15,44,45], however, simple relations based on geometry conditions have been proposed in these studies and absolutely no influence due to the physical properties of the fluids into the predicting modeling has been emphasized in them. Bin [13] was the first who has realized the impact of fluid properties into his empirical predictive modeling:

$$\frac{Q_g}{Q_j} = 0.04 Fr^{0.28} \left(\frac{h_p}{D_N}\right)^{0.4} \tag{13}$$

where Q_g is the downward vertical gas flux density or gas (air) entrainment rate (m^3/s), Q_j is the liquid volumetric rate (m^3/s) and h is the depth of the plume from the impact (m). The above equation has been found agreeable with the data following conditions falling within, $\frac{L_j}{D_N} \leq 100$, $\frac{L_j}{D_N} \geq 10$ and $Fr^{0.28}\left(\frac{L_j}{D_N}\right)^{0.4} \geq 10$. The relation has been observed to agree with an error of about ±20%, compared to experimental values for air entrainment, relating to the present geometrical setup involving air and water.

The present simulated air entrainment ($\frac{Q_g}{Q_j}$) values along with Chanson [39] experimentally measured values and Bin's [13] correlations as well as Euler-Euler simulated values [25], are summarized in Table 4. The VOF simulations have revealed a closer agreement with air entrainment values determined from Bin's correlation and experiments compared to the Euler-Euler simulations. This shows that VOF scheme is effective to trace the air-water interface.

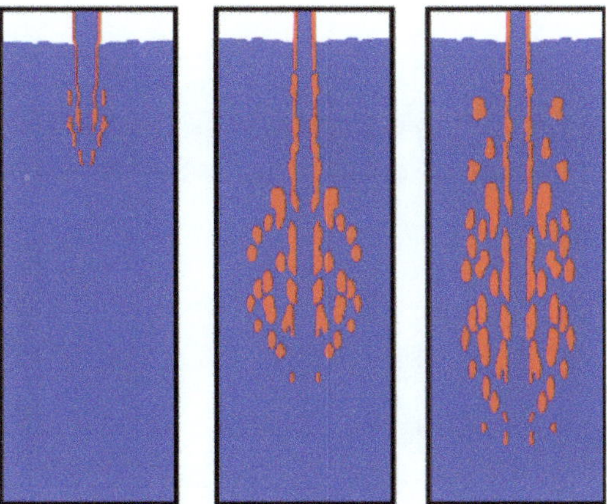

Figure 4. VOF Sequences of bubbly plume.

Moreover, the gas entrainment in the present situation is modeled using the advection (Equation (10)) that provides sufficient protection from the errors in entrainment caused due to the numerical method, whereas, within the region across the developing bubbly plume, the distribution of the bubbles is not influenced by the interfacial forces as only surface tension has been included in the momentum equation. For improvement in the bubble's distribution within the plume's profile across the depth of the water pool, there is a need to improve the modeling by enhancing the interfacial interaction between gas and liquid.

Table 4. Air Entrainment to Water Rate Ratio.

	D_j(m)	u_j (m/s)	L_j (m)	Fr	Q_{air}/Q_w
Bin [13] Correlation					
	0.024	3.5	0.1	7.2	0.12
	0.005	2.54	0.01	11.47	0.104
	0.00683	2.49	0.027	9.62	0.131
Chanson's Values [39]					
	0.024	3.5	0.1	7.2	0.11
	0.00683	2.49	0.027	9.62	0.193
Euler-Euler [25]					
	0.005	2.54	0.01	11.47	0.5
VOF Simulations					
	0.024	3.5	0.1	7.2	0.19
	0.005	2.54	0.01	11.47	0.24
	0.00683	2.49	0.027	9.62	0.26

3.2. Penetration Depth

After nearly four seconds of physical time, the gas plume reaches a steady penetration depth, h_p, below the water free surface. The simulated values for the penetration of bubbly plume can be seen in Figure 5, also recent data for the circular plunging jet from Guyot [43] on the penetration of bubbly plume has been included in this figure. It's interesting to express here the empirical correlation proposed by Bin [13] for plume's penetration (h_p):

$$h_p = 0.42 u_j^{1.25} D_N Q_g^{-0.25} \tag{14}$$

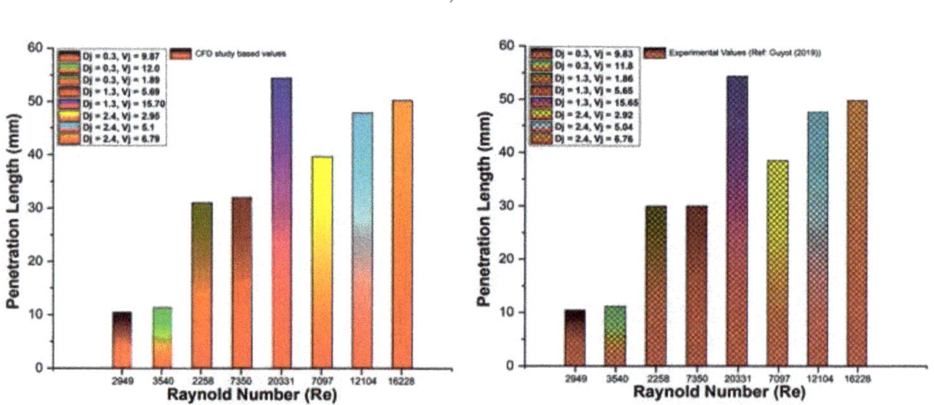

Figure 5. Penetration Length. (**a**) CFD Study-based Values; (**b**) Experimental Values.

The relation has been used to determine the depth of the gas plume as per the sizes of the configuration. The relation is applicable for liquid jet nozzle sizes that vary from 3.9 mm to 12 mm, whereas, and for a pool depth (h) about 0.50 m or less, with the error of ±20%. VOF simulated values for h_p higher than both experimental [39] and predictive values of the above equation. This equation can also be expressed alternatively [18] as:

$$h_p = 2.1 u_j^{0.775} D_N^{0.67} \tag{15}$$

The relation reveals that the bubbly plume is independent of the quantity of air being entrained and it only depends on the jet's velocity and the nozzle size. Table 5 provides a summary of the penetration depth of bubbly plume below the impact surface for a jet velocity of 2.54 m/s and nozzle dia of 5 mm. The VOF computed penetration depth that follows the continuous entrainment regime against the operating condition is 0.3 m, which exceeds both the empirical correlation (Equation (14)) and the experimental values. However, with the Euler–Euler scheme, the plume depth doesn't reach as much as the experiment (Equation (14)).

Table 5. Bubbly Plume Penetration Depth(m) at u_j = 2.54 m/s and D_N = 5 mm.

Experimental	Bin's Correlation	Euler-Euler	VOF Simulation
0.23	0.131	0.10	0.3

3.3. Void Fraction

The instant void fraction values have been averaged over time to obtain a steady mean void fraction over 3000 iterations, which have been averaged again along the circumference. The mean simulated values of void fraction are compared with averaged values of void fraction experimentally measured by Chanson [39], obtained under similar operating conditions as in the simulations described here. The experimental void fractions determined by Chanson [39] have been measured along a horizontal axis that passes through the jet's center. These measurements have been averaged by including results either sides of the jet to curtail the error owing to the measurements. It is known that the flow attains the status of fully developed at $40D_j$ [46], which is much outside the computational domain being used here. Subsequently, the plume's depth such as $0.8D_j$, $1.2D_j$, and $2D_j$ from the impact surface, have been chosen for simulated void fraction profiles across the bubbly plume.

It has been described under the above topics that the bubbles have been entrained through the impact between the liquid jet and the receiving pool. Further from the impact surface, the bubbles move both inside and outside radially, as the two-phase mixture travels deep into the water pool. All this can be seen in a sequence of snaps shown in Figure 4 for the operating conditions involving a liquid jet velocity of 3.5 m/s and a nozzle size of 25 mm.

The simulated peaked void fraction along with the depth ($0.8D_j$, $1.2D_j$, and $2.0D_j$) from the impact surface along with the corresponding experimental measured values of void fraction are presented in Figures 6–8, where the predicted values have been averaged in time. The computed void fraction profiles are obtained for liquid jet velocities of 3.5, 4.1, and 4.4 m/s, to compare both with experimentally measured as well as other simulated void faction profiles.

Figures 9–11 provide plots of the numerically and the experimentally obtained void fraction distributions against liquid jet velocities of 3.5, 4.1 and 4.4 m/s at three depths of 0.8, 1.2, and $2.0D_j$ from the water-free surface, respectively. The bottom profile for all void fraction plots relates to the void fraction distribution obtained at depth $0.8D_j$ from the contact surface. The simulated void fraction values in this plot agree well with the measured values, and the location of the peaked void fraction matches among the measured value and the present VOF simulation. The peaked void fraction for all three downstream locations (0.8, 1.2 & $2.0D_j$) along the depth of the pool from the contact surface is shown in Figures 9–11 for the liquid jet velocity of 3.5, 4.1, and 4.4 m/s.

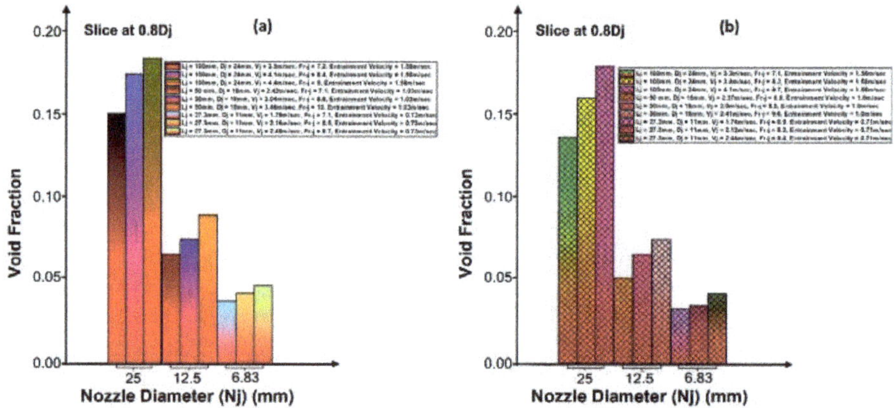

Figure 6. (**a**) VOF Simulated Values; (**b**) Chanson [39] Measured Values for a Peaked Void Fraction at $0.8D_j$ Depth.

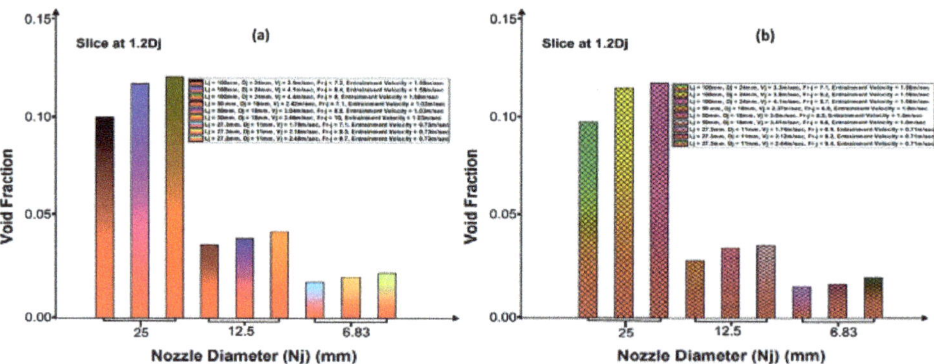

Figure 7. (**a**) VOF Simulated Values; (**b**) Chanson [39] Measured Values for a Peaked Void Fraction at $1.2D_j$ Depth.

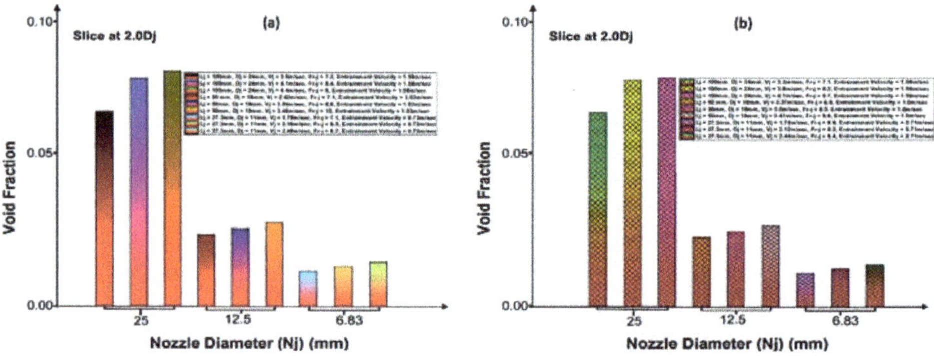

Figure 8. (**a**) VOF Simulated Values; (**b**) Chanson [39] Measured Values for a Peaked Void Fraction at $2.0D_j$ Depth.

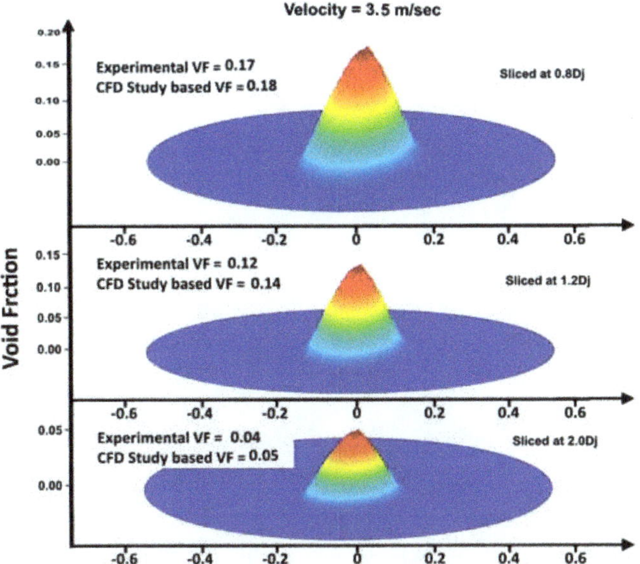

Figure 9. VOF Simulated Void Fraction Distribution at 0.8, 1.2 and 2.0D_j for Jet Velocity of 3.5 m/s.

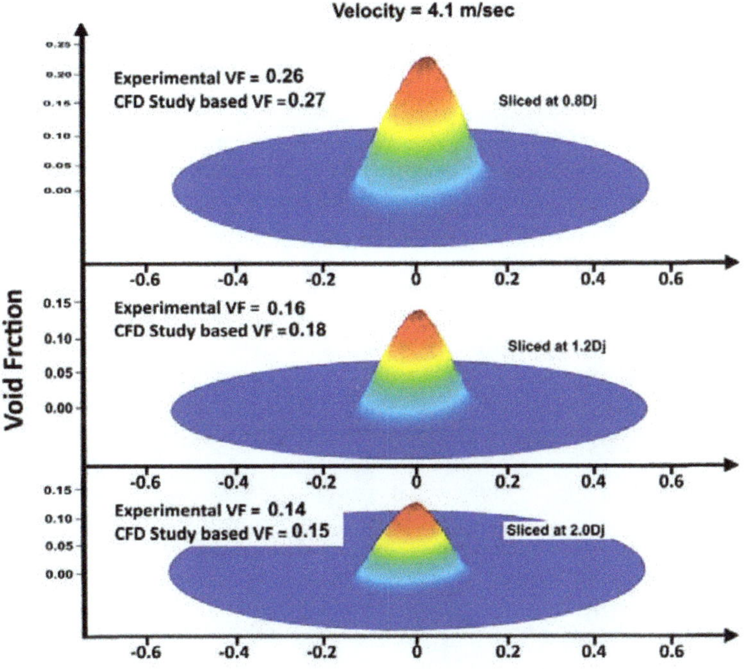

Figure 10. VOF Simulated Void Fraction Distribution at 0.8, 1.2 and 2.0D_j for Jet Velocity of 4.1 m/s.

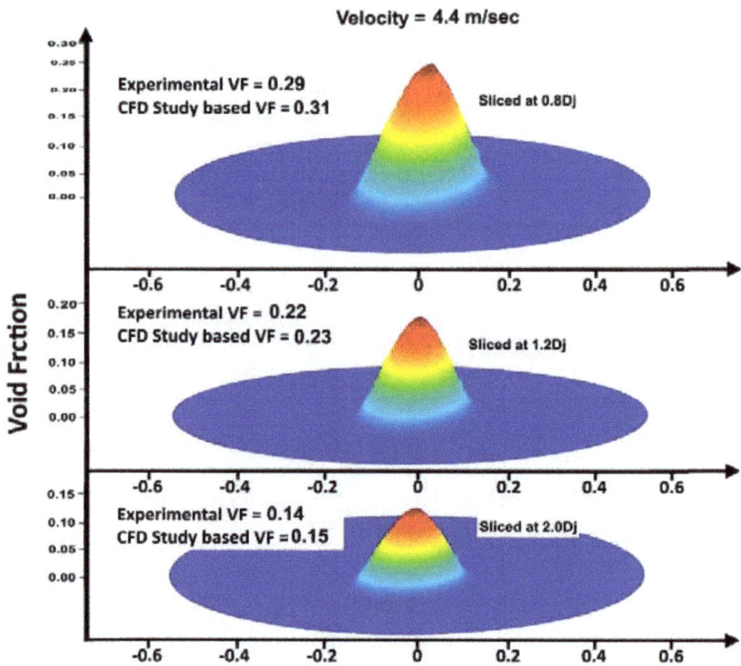

Figure 11. VOF Simulated Void Fraction Distribution at 0.8, 1.2, and 2.0D_j for the Jet Velocity of 4.4 m/s.

Since the simulated bubbly jet is spreading at depth, $0.8D_j$ is small, however, the agreement with measurements suggests that the VOF model accurately depicts the location of air entrainment owing to their sources. Further, it can be seen in these figures that the maximum value of the void fraction agrees well with the measured value, which confirms the accuracy of the VOF model in computing the quantity of air that has been entrained.

The plots further down to lower depths from $0.8D_j$ are presented in Figures 10 and 11, which also indicate good agreement between the experiments and the presented simulations. This indicates that VOF accurately predicts the spreading of bubbly jet and the transport of bubbles which are entrapped by the liquid jet at the impact surface. Transport of bubbles is powered by the liquid velocity along with a balance between turbulent dispersion and drag forces acting on a bubble. Also, the present simulated results make good agreement with the analytical results derived from the simple diffusion scheme adopted by Chanson [39]. Such evidence has been derived by Drew and Passman [47], who have indicated that in circumstances when the dispersed phase momentum under the resulting dominance impact of drag force and turbulent dispersion, the mass conservation should be modeled by a diffusion equation.

Thus, the simulations are obtained at $0.8D_j$, $1.2D_j$, and $2.0D_j$ to enable comparison with measured values of the void fraction by Chanson [39] at the same downstream distance from the impact surface. The comparisons between the time-averaged steady simulated values of void fraction and the measured values are presented in Figures 9–11.

4. Conclusions

CFD simulations of water jet impacting into a still water pool are performed using the VOF formulation through the commercial code FLUENT, whereas tracing of the interface between air and water is conducted by applying a PLIC algorithm. The simulated results are confirmed by comparing them with Chanson [39] experiments through considering uniformity conditions of experimental three jet sizes along with three jet velocities and three jet diameters. The simulations agreed well with experimental data. Focusing VOF simulations on the developing region of the bubbly plume, it has been inferred that the location of peak void fraction in both experiments and the simulations match. In the experiment, this maximum decays exponentially for deeper locations. However, in the present simulations, it remains almost constant as the two phases move with the same velocity. This fact is inherent to the VOF modeling which resolves only one equation. Hence, no slip velocity between the two phases can be modeled. Slight overestimates of air entrainment rate compared to the experiments of Chanson [39] are noticed for all the three cases, however, VOF simulations came close to the experimental values as compared to the empirical correlation and Euler-Euler scheme. This indicates that the entrainment should be regulated by more parameters than just mere geometric similarities. Hence, small scale-models should be investigated keenly to describe the clear picture of the air entrainment.

Funding: This article was funded by the Deanship of Scientific Research (DSR) at King Abdulaziz University, Jeddah. The author, therefore, acknowledges with thanks DSR for technical and financial support.

Conflicts of Interest: The authors declare no conflict of interest.

References

1. Chanson, H. *Air Bubble Entrainment in Free-Surface Turbulent Shear Flows*; Elsevier: Amsterdam, The Netherlands, 1996.
2. Chanson, H.; Aoki, S.; Hoque, A. Bubble entrainment and dispersion in plunging jet flows: Freshwater vs. seawater. *J. Coast. Res.* **2006**, *22*, 664–677. [CrossRef]
3. Harby, K.; Chiva, S.; Muñoz-Cobo, J.-L. An experimental study on bubble entrainment and flow characteristics of vertical plunging water jets. *Exp. Therm. Fluid Sci.* **2014**, *57*, 207–220. [CrossRef]
4. Miwa, S.; Yamamoto, Y.; Chiba, G. Research activities on nuclear reactor physics and thermal-hydraulics in Japan after Fukushima-Daiichi accident. *J. Nucl. Sci. Technol.* **2018**, *55*, 575–598. [CrossRef]
5. Kolani, A.R.; Oguz, H.N.; Prosperetti, A. A New Aeration Device. In Proceedings of the ASME Fluids Engineering Summer Meeting, Washington, DC, USA, 21–25 June 1998.
6. Robison, R. Chicago's waterfalls. *Civ. Eng. N. Y.* **1994**, *64*, 36.
7. Ohkawa, A.; Kusabiraki, D.; Shiokawa, Y.; Sakai, N.; Fujii, M. Flow and oxygen transfer in a plunging water jet system using inclined short nozzles and performance characteristics of its system in aerobic treatment of wastewater. *Biotechnol. Bioeng.* **1986**, *28*, 1845–1856. [CrossRef] [PubMed]
8. Cummings, P.D.; Chanson, H. Air Entrainment in the Developing Flow Region of Plunging Jets—Part 2: Experimental. *ASME J. Fluids Eng.* **1997**, *119*, 603–608. [CrossRef]
9. Jameson, G.L. Bubbly Flows and the Plunging Jet Flotation Column. In Proceedings of the 12th Australasian Fluid Mechanics Conference, Sydney, Australia, 10–15 December 1995.
10. Gupta, A.; Yan, D.S. *Mineral Processing Design Operations: An Introduction*; Elsevier: Amsterdam, The Netherlands, 2016.
11. Leung, S.M.; Little, J.C.; Holst, T.; Love, N.G. Air/Water Oxygen Transfer in a Biological Aerated Filter. *J. Environ. Eng.* **2006**, *132*, 181–189. [CrossRef]
12. Lorenceau, É.; Quéré, D.; Eggers, J. Air Entrainment by a Viscous Jet Plunging into a Bath. *Phys. Rev. Lett.* **2004**, *93*, 254501. [CrossRef]
13. Biń, A.K. Gas entrainment by plunging liquid jets. *Chem. Eng. Sci.* **1993**, *48*, 3585–3630. [CrossRef]
14. Tojo, K.; Naruko, N.; Miyanami, K. Oxygen transfer and liquid mixing characteristics of plunging jet reactors. *Chem. Eng. J.* **1982**, *25*, 107–109. [CrossRef]
15. McKeogh, E.J.; Ervine, D.A. Air entrainment rate and diffusion pattern of plunging liquid jets. *Chem. Eng. Sci.* **1981**, *36*, 1161–1172. [CrossRef]
16. Bonetto, F.; Lahey, R.T., Jr. An experimental study on air carryunder due to a plunging liquid jet. *Int. J. Multiph. Flow* **1993**, *19*, 281–294. [CrossRef]
17. Zhu, Y.; Hasan, N.; Oğuz, A.P. On the mechanism of air entrainment by liquid jets at a free surface. *J. Fluid Mech.* **2000**, *404*, 151–177. [CrossRef]
18. Qu, X.L.; Khezzar, L.; Danciu, D.; Labois, M.; Lakehal, D. Characterization of plunging liquid jets: A combined experimental and numerical investigation. *Int. J. Multiph. Flow* **2011**, *37*, 722–731. [CrossRef]
19. Ma, J.; Oberai, A.A.; Drew, D.A.; Lahey, R.T. A two-way coupled polydispersed two-fluid model for the simulation of air entrainment beneath a plunging liquid jet. *J. Fluids Eng.* **2012**, *134*, 101304. [CrossRef]

20. Yin, Z.; Jia, Q.; Li, Y.; Wang, Y.; Yang, D. Computational Study of a Vertical Plunging Jet into Still Water. *Water* **2018**, *10*, 989. [CrossRef]

21. Ma, J.; Oberai, A.A.; Drew, D.A.; Lahey, R.T.; Moraga, F.J. A quantitative sub-grid air entrainment model for bubbly flows—Plunging jets. *Comput. Fluids* **2010**, *39*, 77–86. [CrossRef]

22. Mirzaii, I.; Passandideh-Fard, M. Modeling free surface flows in presence of an arbitrary moving object. *Int. J. Multiph. Flow* **2012**, *39*, 216–226. [CrossRef]

23. Elahi, R.; Passandideh-Fard, M.; Javanshir, A. Simulation of liquid sloshing in 2D containers using the volume of fluid method. *Ocean Eng.* **2015**, *96*, 226–244. [CrossRef]

24. van Sint Annaland, M.; Deen, N.G.; Kuipers, J.A.M. Numerical simulation of gas bubbles behaviour using a three-dimensional volume of fluid method. *Chem. Eng. Sci.* **2005**, *60*, 2999–3011. [CrossRef]

25. Boualouache, A.; Kendil, F.Z.; Mataoui, A. Numerical assessment of two phase flow modeling using plunging jet configurations. *Chem. Eng. Res. Des.* **2017**, *128*, 248–256. [CrossRef]

26. Shonibare, O.Y.; Wardle, K.E. Numerical Investigation of Vertical Plunging Jet Using a Hybrid Multifluid–VOF Multiphase CFD Solver. *Int. J. Chem. Eng.* **2015**, *2015*, e925639. Available online: https://www.hindawi.com/journals/ijce/2015/925639/ (accessed on 20 March 2020). [CrossRef]

27. Hirt, C.W.; Nichols, B.D. Volume of fluid (VOF) method for the dynamics of free boundaries. *J. Comput. Phys.* **1981**, *39*, 201–225. [CrossRef]

28. Julia, J.E.; Ozar, B.; Jeong, J.-J.; Hibiki, T.; Ishii, M. Flow regime development analysis in adiabatic upward two-phase flow in a vertical annulus. *Int. J. Heat Fluid Flow* **2011**, *32*, 164–175. [CrossRef]

29. Lucas, D.; Rzehak, R.; Krepper, E.; Ziegenhein, T.; Liao, Y.; Kriebitzsch, S.; Apanasevich, P. A strategy for the qualification of multi-fluid approaches for nuclear reactor safety. *Nucl. Eng. Des.* **2016**, *299*, 2–11. [CrossRef]

30. Kothe, D.; Rider, W.; Mosso, S.; Brock, J.; Hochstein, J. Volume tracking of interfaces having surface tension in two and three dimensions. In Proceedings of the 34th Aerospace Sciences Meeting and Exhibit, Reno, NV, USA, 15–18 January 1996.

31. Nguyen, A.V.; Evans, G.M. Computational fluid dynamics modelling of gas jets impinging onto liquid pools. *Appl. Math. Model.* **2006**, *30*, 1472–1484. [CrossRef]

32. Brackbill, J.U.; Kothe, D.B.; Zemach, C. A continuum method for modeling surface tension. *J. Comput. Phys.* **1992**, *100*, 335–354. [CrossRef]

33. Kafka, F.Y.; Dussan, E.B. On the interpretation of dynamic contact angles in capillaries. *J. Fluid Mech.* **1979**, *95*, 539–565. [CrossRef]

34. Boualouache, A.; Zidouni, F.; Mataoui, A. Numerical Visualization of Plunging Water Jet using Volume of Fluid Model. *J. Appl. Fluid Mech.* **2018**, *11*, 95–105. [CrossRef]

35. Gopala, V.R.; van Wachem, B.G. Volume of fluid methods for immiscible-fluid and free-surface flows. *Chem. Eng. J.* **2008**, *141*, 204–221. [CrossRef]

36. Deshpande, S.S.; Trujillo, M.F. Distinguishing features of shallow angle plunging jets. *Phys. Fluids* **2013**, *25*, 082103. [CrossRef]

37. Youngs, D.L. Time-Dependent Multi-Material Flow with Large Fluid Distortion. In *Numerical Methods in Fluid Dynamics*; Morton, K.W., Baines, M.J., Eds.; Academic Press: Cambridge, MA, USA, 1982; pp. 273–285.

38. Launder, B.; Spalding, D. The numerical computation of turbulent flows. *Comput. Method Appl Mech. Eng.* **1974**, *3*, 269–289. [CrossRef]

39. Chanson, H.; Aoki, S.; Hoque, A. Physical modelling and similitude of air bubble entrainment at vertical circular plunging jets. *Chem. Eng. Sci.* **2004**, *59*, 747–758. [CrossRef]

40. Iguchi, M.; Okita, K.; Yamamoto, F. Mean velocity and turbulence characteristics of water flow in the bubble dispersion region induced by plunging water jet. *Int. J. Multiph. Flow* **1998**, *24*, 523–537. [CrossRef]

41. Best Practice Guidelines for the Use of CFD in Nuclear Reactor Safety Applications. 2007. Available online: http://inis.iaea.org/Search/search.aspx?orig_q=RN:44037877 (accessed on 15 July 2020).

42. Versteeg, H.K.; Weeratunge, M. *An Introduction to Computational Fluid Dynamics: The Finite Volume Method*; Pearson: London, UK, 2007.

43. Guyot, G.; Cartellier, A.; Matas, J.-P. Depth of penetration of bubbles entrained by an oscillated plunging water jet. *Chem. Eng. Sci. X* **2019**, *2*, 100017. [CrossRef]

44. Ervine, D.A.; McKeogh, E.; Elsawy, E.M. Effect of turbulence intensity on the rate of air entrainment by plunging water jets. *Proc. Inst. Civ. Eng.* **1980**, *69*, 425–445. [CrossRef]

45. Smith, J.M.; van de Donk, J.A.C. Water Aeration With Plunging Jets. Ph.D. Thesis, Delf University of Technology, Delf, The Netherland, 4 April 1981.
46. Wilcox, D.C. *Turbulence Modeling for CFD*; DCW Industries: New York, NY, USA, 1998; Volume 2.
47. Drew, D.A.; Passman, S.L. *Theory of Multicomponent Fluids*; Springer Science Business Media: New York, NY, USA, 2006; Volume 135.

Publisher's Note: MDPI stays neutral with regard to jurisdictional claims in published maps and institutional affiliations.

MDPI

St. Alban-Anlage 66

4052 Basel

Switzerland

Tel. +41 61 683 77 34

Fax +41 61 302 89 18

www.mdpi.com

ChemEngineering Editorial Office

E-mail: chemengineering@mdpi.com

www.mdpi.com/journal/chemengineering